RECURSION THEORY

LECTURE NOTES IN LOGIC

A Publication of

THE ASSOCIATION FOR SYMBOLIC LOGIC

LECTURE NOTES IN LOGIC 1

Recursion Theory

Joseph R. Shoenfield
Department of Mathematics
Duke University

CRC Press
Taylor & Francis Group
Boca Raton London New York

CRC Press is an imprint of the
Taylor & Francis Group, an **informa** business
AN A K PETERS BOOK

Addresses of the Editors of Lecture Notes in Logic and a Statement of Editorial Policy may be found at the back of this book.

Sales and Customer Service:
A K Peters, Ltd.
888 Worcester Street, Suite 230
Wellesley, Massachusetts 02482, USA
http://www.akpeters.com/

Association for Symbolic Logic:
Sam Buss, Publisher
Department of Mathematics
University of California, San Diego
La Jolla, California 92093-0112, USA
http://www.aslonline.org/

Library of Congress Cataloging-in-Publication Data

Shoenfield, Joseph R. (Joseph Robert), 1927-
Recursion Theory / Joseph R. Shoenfield
p. cm.
Originally published: Berlin ; New York : Springer-Verlag, c1993. (Lecture notes in logic ; 1).
Includes index.
ISBN 1-56881-149-7 (pbk. ; acid-free paper)
1. Recursion theory. I. Title. II. Lecture notes in logic ; 1.

QA9.6.S48 2001
511.3'5-dc21

00-065281

Publisher's note: Joseph R. Shoenfield died on November 15, 2000, while this book was being reprinted. It is published, with the permission of the author, as a photographic reprint of the original edition, for which copyright had reverted to the author. The original was prepared using camera-ready copy prepared by the author. This book was printed by Integrated Book Technology, Troy, NY, on acid-free paper. The cover design is by Richard Hannus, Hannus Design Associates, Boston, Massachusetts.

10 09 08 07 06 5 4 3 2

Introduction

These notes originated in a one semester course given several times at Duke University. Students were generally graduate students in mathematics or computer science. They were expected to have a considerable degree of mathematical maturity. While there were no specific prerequisites, it was desirable that the student has some knowledge of elementary logic and computer programming. To appreciate some of the applications, it was desirable to have at least a slight acquaintance with some other mathematical topics, such as group theory and Borel sets.

My main object was to prepare the student for studying advanced books and journal articles in recursion theory. I therefore attempted to introduce as many as possible of the topics which are of interest in contemporary research without going deeply into any of them. One topic usually treated in the course which did not make its way into these notes is complexity theory; I hope that someone more competent than I will write an introduction to this topic suitable for serious mathematicians.

CONTENTS

1. Computability

Recursion theory is, at least in its initial stages, the theory of computability. In particular, the first task of recursion theory is to give a rigorous mathematical definition of computable.

A computation is a process by which we proceed from some initially given objects by means of a fixed set of rules to obtain some final results. The initially given objects are called inputs; the fixed set of rules is called an algorithm; and the final results are called outputs.

We shall always suppose that there is at most one output; for a computation with k outputs can be thought of as k different computations with one output each. On the other hand, we shall allow any finite number of inputs (including zero).

We shall suppose that each algorithm has a fixed number k of inputs. We do not, ever, require that the algorithm give an output when applied to every k–tuple of inputs. In particular, for some k–tuples of inputs the algorithm may go on computing forever without giving an output.

An algorithm with k inputs computes a function F defined as follows. A k–tuple of inputs $x_1,...,x_k$ is in the domain of F iff the algorithm has an output when applied to the inputs $x_1,...,x_k$; in this case, $F(x_1,...,x_k)$ is that output. A function is computable if there is an algorithm which computes it.

As noted, an algorithm is set of rules for proceeding from the inputs to the output. The algorithm must specify precisely and unambiguously what action is to be taken at each step; and this action must be sufficiently mechanical that it can be done by a suitable computer.

It seems very hard to make these ideas precise. We shall therefore proceed in a different way. We shall give a rigorous definition of a class of functions. It will be clear from the definition that every function in the class is computable. After some study of the class, we shall give arguments to show that every computable function is in the class. If we accept these arguments, we have our rigorous definition of computable.

2. Functions and Relations

We must first decide what inputs and outputs to allow. For the moment, we will take our inputs and outputs to be natural numbers, i.e., non–negative integers. We agree that number means natural number unless otherwise indicated. Lower case Latin letters represent numbers.

We now describe the functions to which the notion of computability applies. Let ω be the set of numbers. For each k, ω^k is the set of k–tuples of numbers. Thus ω^1 is ω, and ω^0 has just one member, the empty tuple. When it is not necessary to specify k, we write $\vec{x}$ for $x_1,\ldots,x_k$.

A k–ary function is a mapping of a subset of ω^k into ω. We agree that a function is always a k–ary function for some k. We use capital Latin letters (usually F, G, and H) for functions.

A k–ary function is total if its domain is all of ω^k. A 0–ary total function is clearly determined by its value at the empty tuple. We identify it with this value, so that a 0–ary total function is just a number. A 1–ary total function is called a real. (This terminology comes from set theory, where reals are often identified with real numbers. It will lead to no confusion, since we never deal with real numbers.)

A common type of algorithm has as output a yes or no answer to some question about the inputs. Since we want our outputs to be numbers, we identify the answer yes with the number 0 and the answer no with the number 1. We now describe the objects computed by such algorithms.

A k–ary relation is a subset of ω^k. We use capital Latin letters (generally P, Q, and R) for relations. If R is a relation, we usually write $R(\vec{x})$ for $\vec{x} \in R$. If R is 2–ary, we may also write $x\ R\ y$ for $R(x,y)$.

A 1–ary relation is simply a set of numbers. We understand set to mean set of numbers; we will use the word class for other kinds of sets. We use A and B for sets.

If R is a k–ary relation, the representing function of R, designated by χ_R, is the k–ary total function defined by

$$\chi_R(\vec{x}) = 0 \qquad \text{if } R(\vec{x}),$$
$$= 1 \qquad \text{otherwise.}$$

A relation R is computable if the function χ_R is computable. We adopt the convention that whenever we attribute to a relation some property usually attributed to a function, we are actually attributing that property to the representing function of the relation.

3. The Basic Machine

To define our class of functions, we introduce a computing machine called the basic machine. It is an idealized machine in that it has infinitely much memory and never makes a mistake. Except for these features, it is about as simple as a computing machine can be.

For each number i, the computing machine has a register $\mathbf{i}$. At each moment, $\mathbf{i}$ contains a number; this number (which has nothing to do with the number i) may change as the computation proceeds.

The machine also has a program holder. During a computation, the program holder contains a program, which is a finite sequence of instructions. If N is the number of instructions in the program, the instructions are numbered 0, 1, ..., $N-1$ (in the order in which they appear in the program). The machine also has a counter, which at each moment contains a number.

To use the machine, we insert a program in the program holder; put any desired numbers in the registers; and start the machine. This causes 0 to be inserted in the counter. The machine then begins executing instructions. At each step, the machine executes the instruction in the program whose number is in the counter at the beginning of the step, provided that there is such an instruction. If at any time the number in the counter is larger than any number of an instruction in the program, then the machine halts. If this never happens, the machine goes on executing instructions forever.

The instructions are of three types. The first type has the format INCREASE $\mathfrak{X}i$. When the machine executes this instruction, it increases the number in $\mathfrak{X}i$ by 1 and increases the number in the counter by 1. The second type has the format DECREASE $\mathfrak{X}i,n$, where n is the number of an instruction in the program. If the machine executes this instruction when the number in $\mathfrak{X}i$ is not 0, it decreases the number in that register by 1 and changes the number in the counter to n. If it executes this instruction when the number in $\mathfrak{X}i$ is 0, it increases the number in the counter by 1. The third type has the format GO TO n, where n is the number of an instruction in the program. When the machine executes this instruction, it changes the number in the counter to n. Note that if $\mathfrak{X}i$ is not mentioned in an instruction, then the instruction does not change the number in $\mathfrak{X}i$ and the number if $\mathfrak{X}i$ does not affect the action of the instruction.

This completes the description of the basic machine. Of course, we have only described the action of the machine, not its physical construction. However, all of the actions of the basic machine can be carried out by a person with pencil and paper and with the program in front of him; he merely keeps track at each step of the number in the counter and the numbers in the registers mentioned in the program.

For each program P for the basic machine and each k, we define an algorithm A_k^P with k inputs. To apply this algorithm to the inputs $x_1,...,x_k$, we start the machine with P in the program holder, $x_1,...,x_k$ in $\mathfrak{X}1,...,\mathfrak{X}k$ respectively, and 0 in all other registers. If the machine eventually halts, the number in $\mathfrak{X}0$ after it halts is the output; otherwise, there is no output. The k-ary function computed by P is the function computed by A_k^P. A k–ary function F is recursive if it is the k–ary function computed by some program for the basic machine. (In accordance with our convention, a relation is recursive iff its representing function is recursive.)

It is clear that every recursive function is computable. It is not at all evident that every computable function is recursive; but, after some study of the recursive functions, we shall argue that this is also the case.

4. Macros

It is tedious to write programs for the basic machine because of the small number of possible instructions. We shall introduce some new instructions and show that they do not enable us to compute any new functions. The idea is familiar to programmers: the use of subroutines, or, as they are often called nowadays, macros.

For each program P for the basic machine, we introduce a new instruction P^*, called the macro of P. When the machine executes this instruction, it begins executing program P (with whatever numbers happen to be in the registers at the time). If this execution never comes to an end, then the execution of P^* is never completed. If the execution of P is completed, the machine changes the number in the counter to 1 more than the number of the instruction P^* and continues executing instructions. The macro machine is obtained from the basic machine by adding all macros of programs for the basic machine as new instructions. We define the notion of a program computing a function for the macro machine as we did for the basic machine.

We say that the program P and P' are equivalent if the following holds. Suppose that we start two machines with P in the program holder of the first machine, P' in the program holder of the second machine, and the same number in $\mathfrak{X}i$ in both machines for all i. Then either both machines will compute forever;

or both machines will halt, and, when they do, the same number will be in $\mathfrak{X}i$ in both machines for every i. Clearly equivalent programs compute the same k–ary function.

4.1. PROPOSITION. Every program for the macro machine is equivalent to a program for the basic machine.

Proof. Let P be a program for the macro machine. For each macro Q^* in P, we replace the instruction Q^* by the sequence of instruction Q. We then number the instructions in the resulting program. Finally, we change each instruction number within an instruction (i.e., each number n in an instruction DECREASE $\mathfrak{X}i,n$ or GO TO n) so that it refers to the same instruction (in P or in one of the Q's) that it did originally. The result is a program P' for the basic machine.

Suppose that we start with two machines as in the definition of equivalent. The machines will perform the same operations until the first executes a macro Q^*. Then both machines begin executing Q. If neither finishes executing Q, we are done. Otherwise, both finish Q with the same number in $\mathfrak{X}i$ in both machines for all i. The number in the counter of the first will be 1 more than the number of Q^*; and the number in the counter of the second will be 1 more than the number of the last instruction in Q. (This is because the execution of Q can only stop by executing the last instruction and having the counter increase by 1.) Thus either both machines will stop, or they will continue performing the same operations. □

4.2. COROLLARY. Every function computed by a program for the macro machine is recursive. □

We now introduce some useful macros. The program

0) DECREASE $\mathfrak{X}i$,0

causes the number in $\mathfrak{X}i$ to be changed to 0. We write the macro of this program as ZERO $\mathfrak{X}i$.

We now want a program to move the number in $\mathfrak{X}i$ into $\mathfrak{X}j$. We could do this by repeatedly decreasing $\mathfrak{X}i$ and increasing $\mathfrak{X}j$; but this would change the number in $\mathfrak{X}i$ to 0. If we want to avoid this, we need another register $\mathfrak{X}k$. We then move the number in $\mathfrak{X}i$ into $\mathfrak{X}j$ and $\mathfrak{X}k$, and then move it back from $\mathfrak{X}k$ to $\mathfrak{X}i$.

In detail, suppose that i, j, and k are distinct. Then the program

0) ZERO $\mathfrak{X}j$,

1) ZERO $\mathfrak{X}k$,

2) GO TO 5,

3) INCREASE $\mathfrak{X}j$,

4) INCREASE $\mathfrak{X}k$,

5) DECREASE $\mathfrak{X}i$,3,

6) GO TO 8,

7) INCREASE $\mathfrak{X}i$,

8) DECREASE $\mathfrak{X}k$,7

causes the number in $\mathfrak{X}i$ to be moved into $\mathfrak{X}j$ without changing the number in $\mathfrak{X}i$. We write the macro of this program as MOVE $\mathfrak{X}i$ TO $\mathfrak{X}j$ USING $\mathfrak{X}k$. (More precisely, this is the macro of the program for the basic machine which is, by 4.1, equivalent to the above program for the macro machine.) Usually we are not interested in $\mathfrak{X}k$; we then write simply MOVE $\mathfrak{X}i$ TO $\mathfrak{X}j$, and understand that $\mathfrak{X}k$ is to be chosen different from all registers mentioned in the program.

Let F be a k-ary recursive function, and let P be a program which computes F. Choose $m > k$ so that P does not mention $\mathfrak{X}i$ for $i \geq m$. Let Q be the program for the macro machine consisting of the following instructions: MOVE $\mathfrak{X}i$ TO $\mathfrak{X}(m+i)$ USING $\mathfrak{X}m$ for $1 \leq i < m$; ZERO $\mathfrak{X}0$; ZERO $\mathfrak{X}i$ for $k < i < m$; P^*; and MOVE $\mathfrak{X}(m+i)$ TO $\mathfrak{X}i$ USING $\mathfrak{X}m$ for $1 \leq i < m$. We leave the reader to check that if Q is executed with $x_1,\ldots,x_k$ in $\mathfrak{X}1,\ldots,\mathfrak{X}k$, then the machine eventually halts iff $F(x_1,\ldots,x_k)$ is defined; and in this case, $F(x_1,\ldots,x_k)$ is in $\mathfrak{X}0$, and the number in $\mathfrak{X}i$ is unchanged unless $i = 0$ or $m \leq i < 2m$.

Now let $i_1,\dots,i_k,j,n_1,\dots,n_m$ be distinct. By changing register numbers in Q, we produce a program Q' with the following property. If Q' is executed with $x_1,\dots,x_k$ in $\mathcal{R}i_1,\dots,\mathcal{R}i_k$, then the machine eventually halts iff $F(x_1,\dots,x_k)$ is defined; and in this case, $F(x_1,\dots,x_k)$ is in $\mathcal{R}j$, and the number in $\mathcal{R}i$ is unchanged unless $i = j$ or i is one of $n_1,\dots n_m$. We write the macro of Q' as

$$F(\mathcal{R}i_1,\dots,\mathcal{R}i_k) \to \mathcal{R}j \text{ USING } \mathcal{R}n_1,\dots,\mathcal{R}n_m.$$

As above, we generally omit USING $\mathcal{R}n_1,\dots,\mathcal{R}n_m$.

5. Closure Properties

We are going to show that the class of recursive functions has certain closure properties; i.e., that certain operations performed on members of the class lead to other members of the class. In later sections, we shall use these results to see that various functions are recursive.

If $1 \leq i \leq k$, we define the total k–ary function I_i^k by $I_i^k(x_1,\dots,x_k) = x_i$. Recall that every number is a 0–ary total function. The successor function Sc is defined by $Sc(x) = x + 1$. The function I_i^k, 0, and Sc are called the initial functions.

5.1. **Proposition.** The initial functions are recursive.

Proof. The function I_i^k is computed by the program

0) MOVE $\mathcal{R}i$ TO $\mathcal{R}0$.

The function 0 is computed by the program

0) ZERO $\mathcal{R}0$.

The function Sc is computed by the program

0) MOVE $\mathcal{R}1$ TO $\mathcal{R}0$;

1) INCREASE $\mathcal{R}0$. □

Because our functions need not be total, we often meet expressions which may be undefined. Thus if F and G are unary, $F(G(x))$ is defined iff x is in the domain of G and $G(x)$ is in the domain of F. Suppose that X and Y are expressions which may be undefined, and which, if they are defined, represent numbers. Then $X \simeq Y$ means that either X and Y are both defined and represent the same number, or X and Y are both undefined. Note that the expression $X \simeq Y$ is always defined.

A class Φ of functions is closed under composition if whenever G, H_1, ..., H_n are in Φ, then so is the F defined by

$$F(\vec{x}) \simeq G(H_1(\vec{x}),...,H_n(\vec{x})).$$

5.2. PROPOSITION. The class of recursive functions is closed under composition.

Proof. Suppose that G, H_1, ..., H_n are recursive, and that F is defined as above, where $\vec{x}$ is $x_1,...,x_k$. Then F is computed by the program

0) $H_1(\mathfrak{R}1,...,\mathfrak{R}k) \rightarrow \mathfrak{R}(k+1)$,

...,

n–1) $H_n(\mathfrak{R}1,...,\mathfrak{R}k) \rightarrow \mathfrak{R}(k+n)$,

n) $G(\mathfrak{R}(k+1),...,\mathfrak{R}(k+n)) \rightarrow \mathfrak{R}0$. □

If G and H are total functions, we may define a total function F by induction on y as follows:

$$F(0,\vec{x}) = G(\vec{x}),$$
$$F(y+1,\vec{x}) = H(F(y,\vec{x}),y,\vec{x}).$$

A class Φ of functions is inductively closed if whenever G and H are total functions in Φ and F is defined as above, then F is in Φ.

5.3. PROPOSITION. The class of recursive functions is inductively closed.

Proof. Let G and H be total recursive functions and let F be defined as above. To improve readability, assume that $\vec{x}$ is just x. Then F is computed by the program

0) $G(\mathfrak{R}2) \rightarrow \mathfrak{R}0$,

1) MOVE $\mathfrak{R}1$ TO $\mathfrak{R}3$,

2) ZERO $\mathfrak{R}1$,

3) GO TO 7,

4) $H(\mathcal{R}0,\mathcal{R}1,\mathcal{R}2) \to \mathcal{R}4$,

5) MOVE $\mathcal{R}4$ to $\mathcal{R}0$,

6) INCREASE $\mathcal{R}1$,

7) DECREASE $\mathcal{R}3$,4.

For suppose that we start the machine with y in $\mathcal{R}1$ and x in $\mathcal{R}2$. After 0), 1), and 2) are executed, we have $F(0,x)$, 0, x, y in $\mathcal{R}0$, $\mathcal{R}1$, $\mathcal{R}2$, $\mathcal{R}3$. If $F(z,x)$, z, x are in $\mathcal{R}0$, $\mathcal{R}1$, $\mathcal{R}2$ and we execute 4), 5), and 6), then $F(z+1,x)$, $z+1$, x are in these registers. This sequence of three steps is repeated y times because of 3) and 7). Hence we finish with $F(y,x)$, y, x in $\mathcal{R}0$, $\mathcal{R}1$, $\mathcal{R}2$. □

Let $X(x)$ be a statement about the number x which is defined for all values of x. We use $\mu xX(x)$ to designate the least x such that $X(x)$. If there is no x such that $X(x)$, then $\mu xX(x)$ is undefined.

A class Φ of functions is <u>μ-closed</u> if whenever R is a relation in Φ (i.e., such that χ_R is in Φ), then the function F defined by $F(\vec{x}) \simeq \mu yR(y,\vec{x})$ is in Φ.

5.4. Proposition. The class of recursive functions is μ-closed.

Proof. If R is a recursive relation and F is defined as above, then F is computed by the program

0)$0 \to \mathcal{R}0$,

1) GO TO 3,

2) INCREASE $\mathcal{R}0$,

3) $\chi_R(\mathcal{R}0,\mathcal{R}1,\ldots,\mathcal{R}k) \to \mathcal{R}(k+1)$,

4) DECREASE $\mathcal{R}(k+1)$,2. □

A class of functions is <u>recursively closed</u> if it contains the initial functions and is closed under composition, inductively closed, and μ-closed. The results of this section are summarized in the proposition:

5.5. Proposition. The class of recursive functions is recursively closed. □

In the next three sections, the only fact about the class of recursive functions that we use will be the fact that it is recursively closed. This will enable us to prove in §9 that the class of recursive functions is the smallest recursively closed class.

6. Definitions of Recursive Functions

We are now going to show that certain kind of definitions of functions and relations always lead to recursive functions and relations. The simplest kind of definition of a function has the form $F(\vec{x}) \simeq$ ____, where ____ is an expression which, if defined, represents a number and which contains only previously defined symbols and variables from the sequence $\vec{x}$. Such a definition is called an explicit definition of F in terms of the symbols which appear in ____ .

6.1. PROPOSITION. If F is defined explicitly in terms of variables and names of recursive functions, then F is recursive.

Proof. We suppose that F is defined by $F(\vec{x}) \simeq$ ____ and use induction on the number of symbols in ____ . If ____ consists of just an x_i, then F is an I_i^k and hence is recursive. Otherwise, ____ is $G(X_1,\ldots,X_n)$ where G is recursive. By the induction hypothesis, we may define a recursive function H_i by $H_i(\vec{x}) \simeq X_i$. Then

$$F(\vec{x}) \simeq G(H_1(\vec{x}),\ldots,H_n(\vec{x}));$$

so F is recursive by 5.2. □

The simplest type of definition of a relation has the form $R(\vec{x}) \leftrightarrow$ ____ where ____ is a statement containing only previously defined symbols and variables from the sequence $\vec{x}$. In order to make sure that this defines a relation, we insist that ____ be defined for all values of $\vec{x}$. We call such a definition an explicit definition of R in terms of whatever symbols appear in ____.

6.2. PROPOSITION. If R is defined explicitly in terms of variables and names of recursive functions and relations, then R is recursive.

Proof. The definition must be $R(\vec{x}) \leftrightarrow Q(X_1,\ldots,X_n)$, where Q is a recursive relation. This may be rewritten as

$$\chi_R(\vec{x}) \simeq \chi_Q(X_1,\dots,X_n).$$

Then R is recursive by 6.1. □

Our intention is to expand the class of symbols which may be used on the right side of explicit definitions of recursive functions and relations. For simplicity, we call such symbols recursive symbols. Thus we have seen that variables and names of recursive functions and relations are recursive symbols. By 5.4, μ is a recursive symbol.

Now we show that the symbols 0, 1, 2, ... are recursive. If 2, say, appears in an explicit definition, we can think of it as a symbol for a 0–ary function applied to zero arguments. Thus we need only show that 2 is a recursive function. Now this function has the explicit definition $2 \simeq Sc(Sc(0))$; since 0 and Sc are recursive, 2 is recursive by 6.1.

6.3. Proposition. Every constant total function is recursive.

Proof. If F is, say, a total function with constant value 2, then F has the explicit definition $F(\vec{x}) \simeq 2$. □

Let ... and $\_\_z\_\_$ be expressions which represent numbers and are defined for all values of their variables. Suppose that ... contains no variable other than $\vec{x}$ and that $\_\_z\_\_$ contains no variable other than $\vec{x}$, y, and z. We can define a total function F by induction as follows:

$$F(0,\vec{x}) = \dots,$$
$$F(y+1,\vec{x}) = \_\_F(y,\vec{x})\_\_.$$

We call this an inductive definition of F in terms of whatever symbols appear in ... and $\_\_z\_\_$.

6.4. Proposition. If F has an inductive definition in terms of recursive symbols, then F is recursive.

Proof. Let F be defined as above. We may define recursive function G and H explicitly by

$$G(\vec{x}) \simeq \ldots,$$

$$H(z,y,\vec{x}) \simeq \underline{\quad z \quad}.$$

Then

$$F(0,\vec{x}) = G(\vec{x}),$$

$$F(y+1,\vec{x}) = H(F(y,\vec{x}),y,\vec{x}).$$

Hence F is recursive by 5.2. □

We have required that our inductive definitions be by induction on the first argument; but this is not essential. Suppose that we have a definition of $F(x,y)$ by induction on y. If $F'(y,x) = F(x,y)$, we can convert that definition into a definition of F' by induction on y. If only recursive symbols are involved, then F' is recursive. But F has the explicit definition $F(x,y) \simeq F'(y,x)$; so F is recursive.

We now give some inductive definitions of some common functions.

$$0 + x = x,$$

$$(y+1) + x = Sc(y+x),$$

$$0 \cdot x = 0,$$

$$(y+1) \cdot x = (y \cdot x) + x,$$

$$x^0 = 1,$$

$$x^{y+1} = x^y \cdot x.$$

Subtraction is not a total function for us, since we do not allow negative numbers. We therefore introduce a modified subtraction $\dot{-}$ defined by $x \dot{-} y = x - y$ if $x \geq y$, $x \dot{-} y = 0$ otherwise. To show that this is recursive, first define a function Pr inductively by

$$Pr(0) = 0,$$

$$Pr(x+1) = x.$$

Then $\dot{-}$ is defined inductively by

$$x \dot{-} 0 = x,$$

$$x \dot{-} (y+1) = Pr(x \dot{-} y).$$

We recall that if X and Y are statements, then $\neg X$ means not X; $X \vee Y$ means X or Y; $X \;\&\; Y$ means X and Y; $X \rightarrow Y$ means if X, then Y; and $X \leftrightarrow Y$ means X iff Y. We call $\neg X$ the negation of X; $X \vee Y$ the disjunction of X and Y; and $X \;\&\; Y$ the conjunction of X and Y. The symbols $\neg$, $\vee$, $\&$, $\rightarrow$, and $\leftrightarrow$ are called propositional connectives. We shall show that all of them are recursive symbols. It is enough to do this for $\neg$ and $\vee$; for we can define $X \;\&\; Y$ to mean $\neg(\neg X \vee \neg Y)$; $X \rightarrow Y$ to mean $\neg X \vee Y$; and $X \leftrightarrow Y$ to mean $(X \rightarrow Y) \;\&\; (Y \rightarrow X)$. Thus we must show that if P and P' are recursive, and Q and R are defined by

$$Q(\vec{x}) \leftrightarrow \neg P(\vec{x}),$$
$$R(\vec{x}) \leftrightarrow P(\vec{x}) \vee P'(\vec{x}),$$

then Q and R are recursive. This follows from the explicit definitions

$$\chi_Q(\vec{x}) \simeq 1 \dot{-} \chi_P(\vec{x}),$$
$$\chi_R(\vec{x}) \simeq \chi_P(\vec{x}) \cdot \chi_{P'}(\vec{x}).$$

We shall now show that the relations $\leq$, $\geq$, $<$, $>$, and $=$ are recursive. (By the relation $\leq$, for example, we mean the relation R defined by $R(x,y) \leftrightarrow x \leq y$.) It is easy to see that

$$\chi_>(x,y) \simeq 1 \dot{-} (x \dot{-} y);$$

so $>$ is recursive. The other relations have the explicit definitions:

$$x \leq y \leftrightarrow \neg(y > x)$$
$$x \geq y \leftrightarrow y \leq x,$$
$$x < y \leftrightarrow \neg(y \leq x),$$
$$x = y \leftrightarrow x \leq y \;\&\; y \leq x.$$

6.5. Proposition. Let $R_1,\ldots,R_n$ be recursive relations such that for every $\vec{x}$, exactly one of $R_1(\vec{x}),\ldots,R_n(\vec{x})$ is true. Let $F_1,\ldots,F_n$ be total recursive functions, and define a total function F by

$$\begin{aligned} F(\vec{x}) &= F_1(\vec{x}) \quad \text{if } R_1(\vec{x}), \\ &= \ldots \\ &= F_n(\vec{x}) \quad \text{if } R_n(\vec{x}). \end{aligned}$$

Then F is recursive.

Proof. We have

$$F(\vec{x}) = F_1(\vec{x})\cdot\chi_{\neg R_1}(\vec{x}) + \ldots + F_n(\vec{x})\cdot\chi_{\neg R_n}(\vec{x}). \ \square$$

We use 6.5 in connection with definitions by cases of functions and relations. For example, suppose that we define F by

$$\begin{aligned} F(x,y) &= x && \text{if } x < y, \\ &= y + 2 && \text{if } y \leq x \ \&\ x = 4, \\ &= 3 && \text{otherwise.} \end{aligned}$$

This comes under 6.5 if we define

$$\begin{aligned} F_1(x,y) &\simeq x, && R_1(x,y) \leftrightarrow x < y, \\ F_2(x,y) &\simeq y + 2, && R_2(x,y) \leftrightarrow y \leq x \ \&\ x = 4, \\ F_3(x,y) &\simeq 3, && R_3(x,y) \leftrightarrow \neg R_1(x,y) \ \&\ \neg R_2(x,y). \end{aligned}$$

A definition by cases of a relation is easily converted into a definition by cases of its representing function. The conclusion is that if we define a total function or relation by cases in terms of recursive symbols, then the function or relation is recursive. We shall consider definition by cases of partial functions in §8.

6.6. **Proposition.** Let F be a total recursive function, and let G be a total function such that $G(\vec{x}) = F(\vec{x})$ for all but a finite number of $\vec{x}$. Then G is recursive.

Proof. Suppose, for example that F is unary and $G(x) = F(x)$ except that $G(3) = 5$ and $G(7) = 2$. Then we can define G by cases as follows:

$$\begin{aligned} G(x) &= 5 && \text{if } x = 3, \\ &= 2 && \text{if } x = 7, \\ &= F(x) && \text{otherwise.} \ \square \end{aligned}$$

6.7. **Corollary.** Every finite relation is recursive.

Proof. If R is finite, $\chi_R(\vec{x}) = 1$ for all but a finite number of $\vec{x}$. $\square$

Recall that a unary relation is simply a set. Then 6.7 shows that every finite set is recursive. The complement of a recursive set is recursive; for the complement of A is $\neg A$. The union and intersection of two recursive sets is recursive; for the union of A and B is $A \vee B$, and the intersection of A and B is A & B.

Recall that $\forall x$ means for all x and $\exists x$ means for some x. We call $\forall x$ a universal quantifier and $\exists x$ an existential quantifier. As we shall see in §13, these are not recursive symbols. We introduce some modified quantifiers, called bounded quantifiers, which are recursive. We let $(\forall x < y)X(x)$ mean that $X(x)$ holds for all x less than y, and let $(\exists x < y)X(x)$ mean that $X(x)$ holds for some x less than y. To see that these are recursive, note that

$$(\forall x < y)X(x) \leftrightarrow \mu x(\neg X(x) \vee x = y) = y,$$

$$(\exists x < y)X(x) \leftrightarrow \mu x(X(x) \vee x = y) < y.$$

To allow us to use bounded quantifiers with $\leq$ instead of $<$, we agree that $(\forall x \leq y)$ means $(\forall x < y+1)$ and similarly for $\exists$.

We summarize the results of this section. If a function or a relation has an explicit definition or an inductive definition or a definition by cases in terms of recursive symbols, then it is recursive. Recursive symbols include variables, names of recursive functions and relations, μ, propositional connectives, and bounded quantifiers. The recursive functions include the initial functions, $+$, $\cdot$, x^y, $\dot{-}$, and all constant total functions. The recursive relations include all finite relations, $<$, $>$, $\leq$, $\geq$, and $=$.

7. Codes

Suppose that we wish to do computations with a class I other than ω as our set of inputs and outputs. One approach is to assign to each member of I a number, called the code of that member, so that different codes are assigned to different members. Given some inputs in I, we first replace each input by its code. We then do a computation with these numerical inputs to obtain a numerical output. If this output is the code of a member of I, that member is the final output of the computation.

We want the first and last steps of the above procedure to be performed according to an algorithm. This means that we should do our coding so that there is an algorithm by which we can find the code of a given member of I, and an algorithm by which, given a number, we can decide if it is the code of a member of I and, if it is, find that member. If this is the case, we say that the coding is effective.

We shall now assign a code to every finite sequence of numbers. Let p_0, p_1, ... be the primes in increasing order. To the finite sequence x_0, ..., x_{k-1} we assign the code

$$p_0^{x_0+1} \cdot \ldots \cdot p_{k-1}^{x_{k-1}+1}.$$

(The empty sequence has the code 1.) The theorem on unique decomposition into prime factors shows that different sequences have different codes. (Note that this would not be true if we omitted the +1's in the exponents.) Since there is an algorithm for decomposing a number into prime factors, this coding is effective.

We shall show that some functions and relations associated with this coding are recursive. To do this, we give definitions of these functions and relations which show, by the results of the last section, that they are recursive. We always take our functions to be total, even when we are only interested in them for certain arguments.

We want to define $Div(x,y)$ to mean that x is divisible by y. The obvious definition is $Div(x,y) \leftrightarrow \exists z(x = y \cdot z)$. This does not show that Div is recursive, since unbounded quantifiers are not recursive. We therefore seek a bound for z. This leads to the definition:

$$Div(x,y) \leftrightarrow (\exists z \leq x)(x = y \cdot z):$$

Next we define the set Pr of primes:

$$Pr(x) \leftrightarrow x > 1 \;\&\; (\forall y < x)(y > 1 \rightarrow \neg Div(x,y)).$$

We then define by induction

$$p_0 = 2,$$
$$p_{i+1} = \mu x(Pr(x) \;\&\; x > p_i).$$

We define

$$exp(x,i) = \mu y(\neg Div(x,p_i^{\,y+1}) \vee x = 0),$$

so that $exp(x,i)$ is the exponent of p_i in the decomposition of x. (The clause $\vee\ x = 0$ is there to make the function total; it makes $exp(0,i) = 0$.)

For each k we define

$$<x_0,\dots,x_{k-1}> = p_0^{\,x_0+1} \cdot \dots \cdot p_{k-1}^{\,x_{k-1}+1}$$

Then $<x_0,\dots,x_{k-1}>$ is the code of the sequence $x_0,\dots,x_{k-1}$.

Next we define

$$lh(x) = \mu i(exp(x,i) = 0),$$
$$(x)_i = exp(x,i) \dot{-} 1.$$

If $x = <x_0,\dots,x_{k-1}>$, then $lh(x) = k$; $(x)_i = x_i$ for $i < k$; and $(x)_i = 0$ for $i \geq k$. Since $z \leq p^z$ for $p > 1$, $exp(x,i) \leq x$. It follows that

(1) $$x \neq 0 \rightarrow (x)_i < x,$$

and hence

(2) $$x_i < <x_1,\dots,x_k>.$$

Since $Div(x,p_i)$ implies $i < p_i \leq x$ for $x > 0$, the set Seq of codes of finite sequences is defined by

$$Seq(x) \leftrightarrow x \neq 0 \;\&\; (\forall i < x)(Div(x,p_i) \rightarrow i < lh(x)).$$

We define $x * y$ so that $<x_1,\dots,x_k> * <y_1,\dots,y_l>$ is $<x_1,\dots,x_k,y_1,\dots,y_l>$:

$$x * y \simeq \mu z(Seq(z) \;\&\; lh(z) = lh(x) + lh(y) \;\&$$
$$(\forall i < lh(x))((z)_i = (x)_i) \;\&\; (\forall i < lh(y))((z)_{lh(x)+i} = (y)_i)).$$

As a first application of these codes, we show how to replace k–ary functions and relations by 1–ary functions and relations. If F is a k–ary function, we define a 1–ary function $<F>$, called the contraction of F, by

$$<F>(x) \simeq F((x)_0,\dots,(x)_{k-1}).$$

We can recover F from $<F>$ by the equation

$$F(x_1,\dots,x_k) \simeq <F>(<x_1,\dots,x_k>).$$

These two equations are called the contraction equations. Considered as explicit definitions, they show that F is recursive iff $<F>$ is recursive. A similar procedure holds for relations; we leave the details to the reader.

As a second application, we consider simultaneous definitions of functions by induction. Suppose we define

$$F_1(0,\vec{x}) = G_1(\vec{x}),$$
$$F_2(0,\vec{x}) = G_2(\vec{x}),$$
$$F_1(y+1,\vec{x}) = H_1(y,\vec{x},F_1(y,\vec{x}),F_2(y,\vec{x})),$$
$$F_2(y+1,\vec{x}) = H_2(y,\vec{x},F_1(y,\vec{x}),F_2(y,\vec{x})),$$

where G_1, G_2, H_1, and H_2 are total and recursive. This defines total functions F_1 and F_2 by induction on y. We shall show that F_1 and F_2 are recursive. It suffices to show that the function F defined by

$$F(y,\vec{x}) = <F_1(y,\vec{x}),F_2(y,\vec{x})>$$

is recursive; for $F_1(y,\vec{x}) = (F(y,\vec{x}))_0$ and $F_2(y,\vec{x}) = (F(y,\vec{x}))_1$. But F has the recursive definition

$$F(0,\vec{x}) = <G_1(\vec{x}),G_2(\vec{x})>,$$
$$F(y+1,\vec{x}) = <H_1(y,\vec{x},(F(y,\vec{x}))_0,(F(y,\vec{x}))_1), H_2(y,\vec{x},(F(y,\vec{x}))_0,(F(y,\vec{x}))_1)>.$$

As a third application, we introduce a more general form of definition by induction in which the value of the function at y depends on all the previous values. If F is a total $(k+1)$–ary function, we define another total $(k+1)$–ary function $\overline{F}$ by

$$\overline{F}(y,\vec{x}) \simeq <F(0,\vec{x}),\dots,F(y-1,\vec{x})>.$$

Thus $\overline{F}(y,\vec{x})$ codes a sequence which gives the values of $F(i,\vec{x})$ for $i < y$. We show that $\overline{F}$ is recursive iff F is recursive. We cannot use the preceding equation as an explicit definition; for we cannot fill in ... until we know the value of the argument y. However, we have the explicit definitions

$$\overline{F}(y,\vec{x}) \simeq \mu z(Seq(z) \;\&\; lh(z) = y \;\&\; (\forall i < y)((z)_i = F(i,\vec{x}))),$$
$$F(y,\vec{x}) \simeq (\overline{F}(y+1,x))_y.$$

Given a total function G, we may define a total function F by induction on y as follows:

$$F(y,\vec{x}) \simeq G(\overline{F}(y,\vec{x}),\, y,\, \vec{x}).$$

We shall show that if G is recursive, then F is recursive. By the above, it is enough to show that $\overline{F}$ is recursive. But $\overline{F}$ has the inductive definition

$$\overline{F}(0,\vec{x}) \simeq \langle\,\rangle,$$
$$\overline{F}(y+1,\vec{x}) = \overline{F}(y,\vec{x}) * \langle G(\overline{F}(y,\vec{x}),\, y\,,\vec{x})\rangle.$$

An inductive definition of this sort is called a <u>course-of-values</u> inductive definition.

8. Indices

We are now going to assign codes to some of the elements in the operation of the basic machine. This will lead to some of the most important theorems of recursion theory.

First, a general remark on coding. Suppose that we want to code the members of I. We may be able to identify each member b of I with a finite sequence $a_1,\ldots,a_k$ of objects which have already been coded. We can then assign to b the code $\langle x_1,\ldots,x_k\rangle$, where x_i is the code of a_i.

We begin by assigning codes to the instructions for the basic machine. We assign the code $\langle 0,i\rangle$ to the instruction INCREASE $\mathfrak{r}i$; the code $\langle 1,i,n\rangle$ to the instruction DECREASE $\mathfrak{r}i,n$; and the code $\langle 2,n\rangle$ to the instruction GO TO n. If P is a program consisting of N instructions with codes $x_1,\ldots,x_N$, we assign the code $\langle x_1,\ldots,x_N\rangle$ to P.

We define

$$Ins(x) \leftrightarrow x = \langle 0,(x)_1\rangle \lor x = \langle 1,(x)_1,(x)_2\rangle \lor x = \langle 2,(x)_1\rangle,$$

$$Prog(x) \leftrightarrow Seq(x) \mathbin{\&} (\forall i < lh(x))(Ins((x)_i) \mathbin{\&} (((x)_i)_0 = 1 \to ((x)_i)_2 < lh(x))$$

$$\mathbin{\&} (((x)_i)_0 = 2 \to ((x)_i)_1 < lh(x))).$$

Thus *Ins* is the set of codes of instructions and *Prog* is the set of codes of programs.

The action of the machine in executing A_k^P (described near the end of §3) with inputs $\mathfrak{x}$ is called the <u>P–computation from $\mathfrak{x}$</u>. If e is the code of P, then P mentions $\mathcal{R}i$ only for $i < e$ by (2) of §8; so the contents of $\mathcal{R}i$ are significant for this computation only for $i < e + k$. At any step in this computation, the <u>register code</u> of the machine is $\langle r_0,r_1,\dots,r_{e+k-1}\rangle$, where r_i is the number in $\mathcal{R}i$. If the computation stops after m steps, it has successive register codes $r_0, \dots, r_m$. We then assign the code $r = \langle r_0,r_1,\dots,r_m\rangle$ to the computation. By (2) of §8, r is larger than any number which appears in a register during the computation. The output of the computation is $U(r)$, where U is the recursive real defined by

$$U(r) = ((r)_{lh(r)\dot{-}1})_0.$$

We define functions *Count* and *Reg* such that if e is the code of P and $x = \langle\mathfrak{x}\rangle$, then after n steps in the P–computation from $\mathfrak{x}$, $Count(e,x,n)$ will be in the counter and $Reg(j,e,x,n)$ will be in $\mathcal{R}j$. We define these functions by a simultaneous induction on n. Writing i for $(e)_{Count(e,n,x)}$,

$$Count(e,x,0) = 0,$$

$$Reg(j,e,x,0) = (x)_{j\dot{-}1} \quad \text{if } j \le lh(x) \mathbin{\&} j \ne 0,$$

$$= 0 \quad \text{otherwise},$$

$$Count(e,x,n+1) = (i)_2 \quad \text{if } (i)_0 = 1 \mathbin{\&} Reg((i)_1,e,x,n) \ne 0,$$

$$= (i)_1 \quad \text{if } (i)_0 = 2,$$

$$= Count(e,x,n) + 1 \quad \text{otherwise},$$

$$Reg(j,e,x,n+1) = Reg(j,e,x,n) + 1 \quad \text{if } (i)_0 = 0 \mathbin{\&} j = (i)_1,$$

$$= Reg(j,e,x,n) \dot{-} 1 \qquad \text{if } (i)_0 = 1 \ \&\ j = (i)_1,$$
$$= Reg(j,e,x,n) \qquad \text{otherwise.}$$

We define

$$Step(e,x,n) \leftrightarrow Count(e,x,n) \geq lh(e) \ \&\ (\forall i < n)(Count(e,x,i) < lh(e)).$$

Then in the above notation, $Step(e,x,n)$ means that the P–computation from $\vec{x}$ takes n steps.

If $\vec{x}$ is a k–tuple, $T_k(e,\vec{x},y)$ mean that e is the code of a program P and y is the code of the P–computation from $\vec{x}$. Thus

$$T_k(e,\vec{x},y) \leftrightarrow Prog(e) \ \&\ Seq(y) \ \&\ Step(e,<\vec{x}>,lh(y)\dot{-}1) \ \&$$
$$(\forall i < lh(y))((y)_i = \overline{Reg}(e+k,e,<\vec{x}>,i)).$$

If e is the code of a program P, $\vec{x}$ is a k–tuple, and A_k^P has an output when applied to the inputs $\vec{x}$, then $\{e\}(\vec{x})$ is that output; otherwise $\{e\}(\vec{x})$ is undefined. Clearly

$$\{e\}(\vec{x}) \simeq U(\mu y T_k(e,\vec{x},y)).$$

This equation is called the Normal Form Theorem.

We say that e is an <u>index</u> of F if $F(\vec{x}) \simeq \{e\}(\vec{x})$ for all $\vec{x}$.

8.1. PROPOSITION. A function is recursive iff it has an index.

Proof. If F is recursive and e is the code of a program which computes F, then e is an index of F. The converse follows from the Normal Form Theorem. □

8.2. ENUMERATION THEOREM (KLEENE). For each k, $\{e\}(x_1,...,x_k)$ is a recursive function of $e,x_1,...,x_k$.

Proof. By the Normal Form Theorem. □

By the Normal Form Theorem, $\{e\}(\vec{x})$ is defined iff there is a y such that $T_k(e,\vec{x},y)$. By the meaning of T_k, this y is then unique; and $\{e\}(\vec{x}) = U(y)$. We call y the <u>computation number</u> of $\{e\}(\vec{x})$. Since y is greater than every number appearing in a register during the P–computation from $\vec{x}$, it is greater than the x_i and $\{e\}(\vec{x})$.

Recall that the results of the last three section depended only on the fact that the class of recursive functions was recursively closed. Thus every recursively closed class contains U and the T_k, and hence, by the Normal Form Theorem, each of the functions $\{e\}$. Hence by 8.1:

8.3. PROPOSITION. The class of recursive functions is the smallest recursively closed class. □

The importance of 8.3 is that it gives us a method of proving that every recursive function has a property P; we have only to show that the class of functions having property P is recursively closed.

We define

$$\{e\}_s(\vec{x}) \simeq U(\mu y(y \leq s \ \& \ T_k(e,\vec{x},y))).$$

Clearly $\{e\}(\vec{x}) \simeq z$ iff $\{e\}_s(\vec{x}) \simeq z$ for some s; in this case, $\{e\}_s(\vec{x}) = z$ for all $s \geq y$, where y is the computation number of $\{e\}(\vec{x})$. Thus $\{e\}_s$ may be thought of as the sth approximation to $\{e\}$. If $\{e\}_s(\vec{x})$ is defined, each x_i is $< s$; so $\{e\}_s$ is a finite function.

8.4. PROPOSITION. The relations P and Q defined by

$$P(e,s,\vec{x},z) \leftrightarrow \{e\}_s(\vec{x}) \simeq z$$

and

$$Q(e,s,\vec{x}) \leftrightarrow \{e\}_s(\vec{x}) \text{ is defined}$$

are recursive.

Proof. We have

$$P(e,s,\vec{x},z) \leftrightarrow (\exists y \leq s)(T_k(e,\vec{x},y) \ \& \ U(y) = z),$$

$$Q(e,s,\vec{x}) \leftrightarrow (\exists y \leq s)\, T_k(e,\vec{x},y). \ \square$$

We shall now use indices to extend 6.5 to partial functions.

8.5. PROPOSITION. Let $R_1,\ldots,R_n$ be recursive relations such that for every $\vec{x}$, exactly one of $R_1(\vec{x}),\ldots,R_n(\vec{x})$ is true. Let $F_1,\ldots,F_n$ be recursive functions, and define a function F by

$$F(\vec{x}) \simeq F_1(\vec{x}) \quad \text{if } R_1(\vec{x}),$$

$$\simeq \ldots$$

$$\simeq F_n(\vec{x}) \quad \text{if } R_n(\vec{x}).$$

Then F is recursive.

  Proof. Let f_i be an index of F_i. Using 6.5, define a total recursive function G by

$$G(\vec{x}) = f_1 \quad \text{if } R_1(\vec{x}),$$

$$= \ldots$$

$$= f_n \quad \text{if } R_n(\vec{x}).$$

We can then define F by $F(\vec{x}) \simeq \{G(\vec{x})\}(\vec{x})$. □

  8.6. PARAMETER THEOREM. If F is a $(k+m)$–ary recursive function, there is a recursive total function S such that

$$(1) \qquad \{S(y_1,\ldots,y_m)\}(\vec{x}) \simeq F(\vec{x},y_1,\ldots,y_m)$$

for all $\vec{x},y_1,\ldots,y_m$.

  Proof. To simplify the notation, we suppose that $m = 1$ and write y for y_1. Suppose that $\vec{x}$ is a k–tuple. Let the program P_y consist of y INCREASE $\mathcal{R}(k+1)$ instructions followed by the macro of a program P which computes F. Then P_y computes the function G defined by $G(\vec{x}) \simeq F(\vec{x},y)$. If we take $S(y)$ to be the code of the program for the basic machine which by 4.1 is equivalent to P_y, then (1) holds.

  It remains to give a definition of S which shows that it is recursive. First we define

$$F(i,y) = \langle 1,(x)_1,(x)_2 + y \rangle \qquad \text{if } (i)_0 = 1,$$

$$= \langle 2,(x)_1 + y \rangle \qquad \text{if } (x)_0 = 2,$$

$$= i \qquad \text{otherwise.}$$

If i is the code of an instruction I, $F(i,y)$ is the code of the instruction obtained from I by increasing every instruction number by y. Let e be the code of P. Then we define $S(y) = S_1(y) * S_2(y)$, where

$$S_1(y) = \mu z(Seq(z) \ \& \ lh(z) = y \ \& \ (\forall i < y)((z)_i = \langle 0,k+1 \rangle))$$

and

$$S_2(y) = \mu z(Seq(z) \mathbin{\&} lh(z) = lh(e) \mathbin{\&} (\forall i < lh(z))((z)_i = F((e)_i, y))). \ \square$$

An implicit definition of a function F has the form $F(\vec{x}) \simeq$ ____ where now ____ may contain F as well as previously defined symbols and the variables in $\vec{x}$. Of course, this is not really a definition of F; it merely tells us to search for an F which satisfies the equation $F(\vec{x}) \simeq$ ____. Thus $F(\vec{x}) \simeq F(\vec{x})$ is satisfied by every F, and $F(\vec{x}) \simeq F(\vec{x}) + 1$ is satisfied only by the function whose domain is the empty set.

Let us rewrite our implicit definition as $F(\vec{x}) \simeq G(F, \vec{x})$. We would like to show that if G is recursive, then this has at least one recursive solution. Unfortunately, G is not a function in our sense because of the argument F. We therefore replace F as an argument to G by an index of F.

8.7. RECURSION THEOREM (KLEENE). If G is recursive, there is a recursive function F with an index f such that $F(\vec{x}) \simeq G(\vec{x}, f)$ for all $\vec{x}$.

Proof. Since $\{y\}(\vec{x}, y)$ is a recursive function of $\vec{x}, y$, the Parameter Theorem implies that there is a recursive total function S such that $\{S(y)\}(\vec{x}) \simeq \{y\}(\vec{x}, y)$ for all $\vec{x}$ and y. Define a recursive function H by

$$H(\vec{x}, y) \simeq G(\vec{x}, S(y))$$

and let h be an index of H. Let $F = \{S(h)\}$, $f = S(h)$, so that F has index f. Then

$$F(\vec{x}) \simeq \{S(h)\}(\vec{x}) \simeq \{h\}(\vec{x}, h) \simeq H(\vec{x}, h) \simeq G(\vec{x}, S(h)) \simeq G(\vec{x}, f). \ \square$$

The Recursion Theorem is often useful for showing that functions are recursive. For example, suppose that we define

$$\text{(2)} \qquad F(0, x) \simeq G(x),$$
$$F(y+1, x) \simeq H(F(y, 2x), y, x),$$

where G and H are total recursive functions. This is a legitimate definition by induction on y; it uniquely defines a function F, which is total. Our previous methods will not show that F is recursive, since they allow $F(y,x)$ but not $F(y,2x)$ on the right side. We define a recursive L by

$$L(y,x,f) \simeq G(x) \qquad \text{if } y=0,$$
$$\simeq H(\{f\}(y \dot{-} 1,2x),\ y \dot{-} 1,\ x) \qquad \text{otherwise.}$$

Then we use the Recursion Theorem to obtain a recursive F with an index f such that $F(y,x) \simeq L(y,x,f)$. Clearly F satisfies (2); so the function defined by (2) is recursive.

9. Church's Thesis

We have already remarked that it is clear that every recursive function is computable. The statement that every computable function is recursive is known as Church's Thesis. It was proposed by Church about 1934 and has since come to be accepted by almost all logicians. We shall discuss the reasons for this.

Since the notion of a computable function has not been defined precisely, it may seem that it is impossible to give a proof of Church's Thesis. However, this is not necessarily the case. We understand the notion of a computable function well enough to make some statements about it. In other words, we can write down some axioms about computable functions which most people would agree are evidently true. It might be possible to prove Church's Thesis from such axioms. However, despite strenuous efforts, no one has succeeded in doing this (although some interesting partial results have been obtained).

We are thus reduced to trying to give arguments for Church's Thesis which seem to be convincing. We shall briefly examine these arguments.

The first argument is that all the computable functions which have been produced have been shown to be recursive, using, for the most part, the techniques which we have already described. Moreover, all the known techniques for producing new computable functions from old ones (such as definition by induction or by cases) have been shown to lead from recursive functions to recursive functions.

Another argument comes from various attempts to define computable precisely. We have seen two of these: the definition by means of the basic machine and the definition by means of recursively closed classes (see Proposition 8.3). There are many others, some similar to these two and some quite different. In every case, it has been proved that the class of functions so defined is exactly the class of recursive functions. This at least shows that the class of recursive functions is a very natural class; and it is hard to see why this should be so unless it is indeed the class of computable functions.

Now let us consider how we might generalize the basic machine to produce a new computable function. Since the registers can contain an arbitrarily large finite number of arbitrary numbers, and since so much information can be coded in a single number, it seems pointless to increase the amount of memory. We therefore need to add new instructions. For each new instruction, we must, of course, have an algorithm for executing that instruction. Essentially that means that the functions *Reg* and *Count* of §8, when extended to allow for the new instructions, must be computable. Since these function control just one step in the computation, they should be relatively simple computable functions. We might therefore agree (perhaps on the basis of the above arguments) that *Reg* and *Count* must be recursive. But then we could repeat the remaining definitions of §8 and show that every function computed by the new machine is recursive.

We could elaborate on all of these arguments. However, most people become convinced of Church's Thesis only by a detailed study of recursion theory. The most convincing argument is that all of the results of recursion theory become quite reasonable (or even obvious) when recursive is replaced by computable.

We shall henceforth accept Church's Thesis. It will be used in two ways. First, there are many natural and interesting questions about computable functions. We use Church's Thesis to convert these into precise mathematical questions. Here there seems to be no way to proceed without Church's Thesis. Second, we sometimes use Church's Thesis to prove a function is recursive by observing that it is computable and using Church's Thesis to conclude that it is recursive. This type of use is non–essential; we could always use the methods we have developed to prove that the function is recursive. One of the best ways to convince oneself of Church's Thesis is to examine many such examples and see that in every case the function turns out to be recursive.

10. Word Problems

The initial aim of recursion theory was to show that certain problems of the form "Find an algorithm by which ..." were unsolvable. We shall give a few examples of such problems.

Let us first see how to obtain a non–recursive real F. By 8.1, it is enough to make F different from each $\{e\}$. We shall do this by making it different from $\{e\}$ at the argument e. (This idea, known as the diagonal argument, was used first by Cantor to prove that the set of real numbers is uncountable.) In more detail, we define

$$F(e) \simeq \{e\}(e) + 1 \qquad \text{if } \{e\}(e) \text{ is defined,}$$
$$\simeq 0 \qquad \text{otherwise.}$$

It follows from this construction that the set P defined by

$$P(e) \leftrightarrow \{e\}(e) \text{ is defined}$$

is not recursive; for otherwise, the definition of F would be a definition by cases using only recursive symbols, and hence F would be recursive. Thus, using Church's Thesis, we have our first unsolvable problem: find an algorithm for deciding if $\{e\}(e)$ is defined.

Consider the following problem, called the halting problem: Find an algorithm by which, given a program P and an x, we can decide if the computation of P from x halts. Let P be a program which computes the recursive function F defined by $F(e) \simeq \{e\}(e)$. Then the machine halts with program P and input e iff $\{e\}(e)$ is defined. It follows that the halting program is unsolvable, even for this one program P.

To introduce our next problem, we need a few definitions. An alphabet is a finite sequence of symbols. If Ω is an alphabet, an Ω–word is a finite sequence of Ω–symbols. An Ω–production is an expression $X \to Y$, where X and Y are Ω–words. An Ω–process is a finite set of Ω–productions. We usually suppose Ω is fixed and omit the prefixes Ω.

If X and Y are words and P is a production $Z \to V$, then $X \to_P Y$ means that Y results from X by replacing an occurrence of Z in X by V. If W is a process, $X \to_W Y$ means that $X \to_P Y$ for some production P in W; and $X \Rightarrow_W Y$ means that there are words $Z_1, \ldots, Z_k$ such that Z_1 is X, Z_k is Y, and $Z_i \to_W Z_{i+1}$ for $1 \leq i < k$.

The word problem for an alphabet Ω is the following: Find an algorithm by which, given an Ω–process W and Ω–words X and Y, we can decide if $X \Rightarrow_W Y$. We shall show that this problem is unsolvable, even for a particular choice of W and Y.

Let P be a program for which the halting problem is unsolvable. We shall construct a process W. We use a, b, c, d, and e, possibly with subscripts, for symbols of our alphabet. We use a^r for the expression consisting of r occurrences of a. Let N be the number of instruction in P and let $M > 1$ so that $i < M$ for every i such that $\mathfrak{R}i$ is mentioned in P. If r_i is the number in $\mathfrak{R}i$ and n is the number in the counter, the word

$$\mathrm{b}\,\mathrm{c}_n\,\mathrm{b}_0\,\mathrm{a}^{r_0}\,\mathrm{b}_1\,\mathrm{a}^{r_1}\ldots\mathrm{b}_{M-1}\mathrm{a}^{r_{M-1}}\,\mathrm{b}_M$$

is called the status word. Thus if we do the computation of P from x, the initial status word is $\mathrm{bc}_0\mathrm{b}_0\mathrm{b}_1\mathrm{a}^x\mathrm{b}_2\ldots\mathrm{b}_M$, which we write as Z_x. We construct W so that $Z_x \Rightarrow_W \mathrm{bc}_N$ iff the computation of P from x halts. This will imply that there is no algorithm by which, given x, we can decide if $Z_x \Rightarrow_W \mathrm{bc}_N$.

Suppose the machine is executing P. If X is a status word beginning with bc_n, where $n < N$, then there is a next status word Y. We shall put productions in W to insure that $X \Rightarrow_W Y$.

Suppose first that instruction n in P is INCREASE $\mathfrak{X}i$. We put into W the productions $\mathrm{c}_n\mathrm{b}_j \rightarrow \mathrm{b}_j\mathrm{c}_n$ for $j < i$ and the production $\mathrm{c}_n\mathrm{a} \rightarrow \mathrm{ac}_n$. Applying these productions to X enables us to move the c_n until it stands just before b_i. We also put $\mathrm{c}_n\mathrm{b}_i \rightarrow \mathrm{d}_n\mathrm{b}_i\mathrm{a}$ into W; this enables us to increase the number in $\mathfrak{X}i$ by 1 while changing c_n to d_n. The productions $\mathrm{ad}_n \rightarrow \mathrm{d}_n\mathrm{a}$ and $\mathrm{b}_j\mathrm{d}_n \rightarrow \mathrm{d}_n\mathrm{b}_j$ for $j < i$ enable us to move the d_n until it stands just after b. The production $\mathrm{bd}_n \rightarrow \mathrm{bc}_{n+1}$ then gives Y.

Now suppose that instruction n in P is DECREASE $\mathfrak{X}i,m$. Just as above, the productions $\mathrm{c}_n\mathrm{b}_j \rightarrow \mathrm{b}_j\mathrm{c}_n$ for $j < i$, $\mathrm{c}_n\mathrm{a} \rightarrow \mathrm{ac}_n$, $\mathrm{c}_n\mathrm{b}_i\mathrm{a} \rightarrow \mathrm{d}_n\mathrm{b}_i$, $\mathrm{ad}_n \rightarrow \mathrm{d}_n\mathrm{a}$, $\mathrm{b}_j\mathrm{d}_n \rightarrow \mathrm{d}_n\mathrm{b}_j$ for $j < i$, and $\mathrm{bd}_n \rightarrow \mathrm{bc}_m$ take care of the case in which the number in $\mathfrak{X}i$ is not 0. If it is 0, the above productions again move c_n until it is just before b_i. Then $\mathrm{c}_n\mathrm{b}_i\mathrm{b}_{i+1} \rightarrow \mathrm{e}_n\mathrm{b}_i\mathrm{b}_{i+1}$ changes c_n to e_n; $\mathrm{ae}_n \rightarrow \mathrm{e}_n\mathrm{a}$ and $\mathrm{b}_j\mathrm{e}_n \rightarrow \mathrm{e}_n\mathrm{b}_j$ for $j < i$ bring e_n to just after b; and $\mathrm{be}_n \rightarrow \mathrm{bc}_{n+1}$ gives Y.

If instruction n is GO TO m, then $\mathrm{bc}_n \rightarrow \mathrm{bc}_m$ changes X to Y.

We also add the productions $\mathrm{c}_N\mathrm{b}_i \rightarrow \mathrm{c}_N$ for all i and the production $\mathrm{c}_N\mathrm{a} \rightarrow \mathrm{c}_N$; these enable us to convert any status word beginning with bc_N to bc_N. Hence if the computation of P from x halts, then $Z_x \Rightarrow_W \mathrm{bc}_N$.

A word is <u>special</u> if it contains exactly one occurrence of the c_i, d_i, and e_i symbols. Every status word is special. Moreover, if X is special and $X \rightarrow_W Y$, then Y is special. It is easily checked that if X is special, there is at most one Y such that $X \rightarrow_W Y$.

Now suppose that the computation of P from x never halts. Then there is an infinite sequence X_0, X_1, ... beginning with Z_x such that $X_i \rightarrow_W X_{i+1}$ for all i. The remarks of the previous paragraph then show that the X_i are the only words

X such that $Z_x \Rightarrow_W X$. Hence we cannot have $Z_x \Rightarrow_W \mathrm{bc}_N$; for there is no word V such that $\mathrm{bc}_N \rightarrow_W V$. This completes our proof that the word problem is unsolvable.

A process W is symmetric if whenever it contains $X \rightarrow Y$ it also contains $Y \rightarrow X$. We will show that the word problem is unsolvable even for symmetric processes.

Let W be the process constructed above. Let W' be the symmetric process obtained from W by adding the production $Y \rightarrow X$ for every production $X \rightarrow Y$ in W. We shall show that $Z_x \Rightarrow_{W'} \mathrm{bc}_N$ iff $Z_x \Rightarrow_W \mathrm{bc}_N$; it will then follow that the word problem for W' is unsolvable.

It is enough to show that $Z_x \Rightarrow_{W'} \mathrm{bc}_N$ implies $Z_x \Rightarrow_W \mathrm{bc}_N$. Let $X_1, \ldots, X_k$ be a sequence of the minimum length such that X_1 is Z_x, X_k is bc_N, and $X_i \rightarrow_{W'} X_{i+1}$ for $1 \leq i < k$. Since X_1 is special, it follows by induction on i that X_i is special. If $X_i \rightarrow_W X_{i+1}$ holds for all $i < k$, we are done. If not, pick the largest i such that this is false. Then $X_{i+1} \rightarrow_W X_i$. It follows that X_{i+1} is not bc_N; so $i+1 < k$. By choice of i, $X_{i+1} \rightarrow_W X_{i+2}$. Since X_{i+1} is special, it follows that $X_i = X_{i+2}$. But this means that we could omit X_i and X_{i+1} from our sequence, contradicting the choice of that sequence.

Symmetric processes are interesting because of their relation to semigroups. A semigroup is a class S with a binary operation $\cdot$ such that the associative law holds (i.e., $(x \cdot y) \cdot z = x \cdot (y \cdot z)$ for all $x, y, z \in S$) and there is a unit element (i.e., an $e \in S$ such that $e \cdot x = x \cdot e = x$ for all $x \in S$).

Let Ω be an alphabet. An Ω–semigroup consists of a semigroup S and an element x_{a} of S for every symbol a in Ω. We think of the symbol a as designating the element x_{a}. More generally, the word $\mathrm{a}_1 \ldots \mathrm{a}_k$ designates the element $x_{\mathrm{a}_1} \cdot \ldots \cdot x_{\mathrm{a}_k}$. (Note that no parentheses are needed because of the associative law.)

An <u>Ω–relation</u> is an expression $X = Y$ where X and Y are Ω–words. Then if S is an Ω–semigroup, $X = Y$ is either true or false in S. Now let R be a finite set of Ω–relations and let K be an Ω–relation. Then $R \Rightarrow K$ means that K is true in every Ω–semigroup in which all of the relations in R are true. The <u>word problem for Ω–semigroups</u> is to find an algorithm by which, given R and K, we can decide if $R \Rightarrow K$.

We shall show that the word problem for Ω–semigroups is unsolvable. (This was proved independently by Post and Markov.) Let W' be the symmetric process constructed above. Let R consist of the relations $X = Y$ such that $X \rightarrow Y$ is in W' (and hence $Y \rightarrow X$ is in W'). We shall show that $X \Rightarrow_{W'} Y$ iff $R \Rightarrow X = Y$. Hence the word problem for Ω–semigroups is unsolvable even for this particular R.

Clearly $X \Rightarrow_{W'} Y$ implies $R \Rightarrow X = Y$. To prove the implication in the other direction, we construct an Ω–semigroup. First note that the relation $X \Rightarrow_{W'} Y$ between X and Y is an equivalence relation on the class of Ω–words; this follows from the fact that W' is symmetric. Let X^* be the equivalence class of X. Let S be the set of all these equivalence classes; and define a binary operation $\cdot$ on S by $X^* \cdot Y^* = (XY)^*$ (where XY is X followed by Y). A little thought shows that $(XY)^*$ depends only on the equivalence classes X^* and Y^*; so our definition makes sense. It is easy to see that S is then a semigroup; the unit element is the equivalence class of the empty word.

We make S into an Ω–semigroup by letting the symbol a represent a^*; the word X then represents X^*. If $X = Y$ is in R, then X and Y are equivalent; so $X^* = Y^*$; so $X = Y$ is true in S. It follows that if $R \Rightarrow X = Y$, then $X = Y$ is true in S and hence $X \Rightarrow_{W'} Y$. This completes our proof.

11. Undecidable Theories

We shall see how some problems of the following type can be shown to be unsolvable: find an algorithm by which we can decide if a given sentence is derivable from a system of axioms. The approach given here is due to Tarski. Although we include all necessary definitions, the reader will probably need some familiarity with first–order theories and their models to see what is going on.

A language is a finite set L of symbols, each of which is designated as either a k–ary relation symbol or a k–ary function symbol for some k. A 0–ary function symbol is called a constant. A stucture M for L then consists of the following: (a) a non–empty class $|M|$, called the universe of M; (b) for each k–ary relation symbol R in L, a subset R_M of $|M|^k$; (c) for each k–ary function symbol F in L, a mapping F_M of $|M|^k$ into $|M|$. Members of $|M|$ are called individuals of M. Note that if c is a constant, then c_M is a mapping of $|M|^0$ onto $|M|$ and hence can be identified with an individual of M.

Let L be a language. We introduce an infinite sequence of symbols called variables. We use x, y, and z for variables. We introduce some expressions, called terms, by the following rules: (a) a variable is a term; (b) if F is a k–ary function symbol and $t_1,\ldots,t_k$ are terms, then $F(t_1,\ldots,t_k)$ is a term. We use s and t for terms.

An atomic formula is either an expression of the form $s = t$, or an expression of the form $R(t_1,\ldots,t_k)$, where R is a k–ary relation symbol. The formulas are obtained by the rules: (a) an atomic formula is a formula; (b) if ϕ and ψ are formulas, then $\neg\phi$, $\phi \vee \psi$, $\phi \,\&\, \psi$, $\phi \to \psi$, and $\phi \leftrightarrow \psi$ are formulas; (c) if ϕ is a formula, then $\exists x\phi$ and $\forall x\phi$ are formulas.

An occurrence of x in a formula ϕ is bound if it occurs in a part of ϕ of the form $\exists x\psi$ or $\forall x\psi$; otherwise it is free. A sentence is a formula with no free variables.

Let M be a structure for L. If t is a term and each variable in t represents a particular individual of M, then t represents a particular individual of M. If ϕ is a formula and each free variable of ϕ represents a particular individual of M, the ϕ is either true or false in M. In particular, if ϕ is a sentence, the ϕ is either true or false in M.

A theory consists of a language L and a set of sentences in L; these sentences are called the axioms of the theory. If T is a theory, a model of T is a structure M for the language of T such that every axiom of T is true in M. A theorem of T is a sentence of the language of T which is true in every model of T. The decision problem for a theory T is the following: find an algorithm for deciding if a given sentence of the language of T is a theorem of T. If this problem is unsolvable, we say that T is undecidable.

A theory T' is a finite extension of a theory T if T' is obtained from T by adding a finite number of new axioms.

11.1. PROPOSITION. If T' is a finite extension of T and T' is undecidable, then T is undecidable.

Proof. Let ψ be the conjunction of the axioms added to T to get T'. For every ϕ, ϕ is a theorem of T' iff $\psi \to \phi$ is a theorem of T. Hence a solution to the decision problem for T would give a solution to the decision problem for T'. □

A structure M is strongly undecidable if every theory having M as a model is undecidable. We shall construct such a structure.

Let L be the language whose only member is the binary function symbol $\cdot$. Then every semigroup is a structure for L. Now let Ω be an alphabet, and let L_Ω be obtained from L by adding all the members of Ω as new constants. Then every Ω–semigroup is a structure for L_Ω. We identify an Ω–word $a_1 \ldots a_n$ with the term $a_1 \cdot \ldots \cdot a_n$ of L_Ω; so every Ω–relation is an atomic formula of L_Ω. (Strictly speaking, we should tell how to insert parentheses in the term $a_1 \cdot \ldots \cdot a_n$; but the way in which this is done is immaterial because of the associative law.)

Now let R and S be as at the end of §10. Then R is a finite set of Ω–relations such that there is no algorithm for deciding, given X and Y, whether

$R \Rightarrow X = Y$; and S is an Ω–semigroup such that for all X and Y, $X = Y$ holds in S iff $R \Rightarrow X = Y$. We consider S as a structure for L_Ω. We shall show that S is strongly undecidable.

Let T have S as a model. By 11.1, it is sufficient to prove that some finite extension T' of T is undecidable. We obtain T' from T by adding as new axioms the two semigroup axioms,

$$\forall x \forall y \forall z((x \cdot y) \cdot z = x \cdot (y \cdot z))$$

and

$$\exists y \forall x(y \cdot x = x \;\&\; x \cdot y = x),$$

and the members of R. Then S is a model of T'.

We show that $X = Y$ is a theorem of T' iff $R \Rightarrow X = Y$; this will clearly show that T' is undecidable. If $X = Y$ is a theorem of T', then it is true in S; so $R \Rightarrow X = Y$. Now let $R \Rightarrow X = Y$; we must show that $X = Y$ is true in every model S' of T'. But S' is clearly an Ω–semigroup in which every member of R is true; so $X = Y$ is true in S'.

If we consider S as a structure for L, we have a new structure, and we cannot immediately conclude that it is strongly undecidable. However, we shall show that this is the case.

If every symbol of the language of M is a symbol of the language of M' and $F_M = F_{M'}$ for every symbol F of the language of M, we say that M' is an expansion of M. If, in addition, every symbol of M' which is not a symbol of M is a constant, we say that M' is an inessential expansion of M.

11.2. PROPOSITION. If M' is an inessential expansion of M and M' is strongly undecidable, then M is strongly undecidable.

Proof. Let M be a model of T. Obtain T' from T by adding the constants of M' not constants of M as new symbols (but adding no new axioms). Clearly M' is a model of T'; so T' is undecidable. It therefore suffices to show that a solution of the decision problem for T would give a solution of the decision problem for T'.

Let ϕ be a sentence of T'. Then there is a formula ψ of T such that ϕ is obtained from ψ by replacing the free occurrences of $x_1,...,x_k$ by $c_1,...,c_k$, where $c_1,...,c_k$ are new constants. To prove the desired result, it suffices to show that ϕ is a theorem of T' iff $\forall x_1...\forall x_k\psi$ is a theorem of T. If $\forall x_1...\forall x_k\psi$ is a theorem of T, it is a theorem of T'; so ϕ is a theorem of T'. Suppose ϕ is a theorem of T'. We must show that if N is a model of T, then ψ is true in N when $x_1,...,x_k$ represent individuals of N. Expand N to N' by letting $(c_i)_{N'}$ be the individual represented by x_i. Then N' is a model of T'; so ϕ is true in N'; so ψ is true in N when $x_1,...,x_k$ represet $(c_1)_{N'},...,(c_k)_{N'}$. □

It follows that S considered as a structure for L is strongly undecidable. This gives two interesting undecidable theories. The first has the language L and the two semigroup axioms as its axioms; it is called the <u>theory of semigroups</u>. The second has the language of L and no axioms.

We now introduce our main method for obtaining new strongly undecidable structures from old ones. Let ϕ be a formula of M, and let $x_1,...,x_k$ include all the variables free in ϕ. Let R be the k-ary relation in $|M|$ defined by: $R(a_1,...,a_k)$ holds iff ϕ is true in M when $x_1,...,x_k$ represent $a_1,...,a_k$ respectively. We say that R is the relation <u>defined</u> in M by ϕ using $x_1,...,x_k$. Note that if R is defined in M by ϕ using $x_1,...,x_k$, then for any $y_1,...,y_k$, R is defined in M by some ϕ' using $y_1,...,y_k$. A relation is <u>definable</u> in M if it is defined in M by some formula using some sequence of variables.

Let M and M' be structures such that $|M| \subseteq |M'|$. Then M is <u>definable</u> in M' if $|M|$ is definable in M'; R_M is definable in M' for every relation symbol R of M; and the graph of F_M is definable in M' for every function symbol F of M. We shall show that if this is the case and M is strongly undecidable, then M' is strongly undecidable.

A formula of M is <u>special</u> if every atomic formula occuring in ϕ is either of the form $x = y$ or $R(x_1,...,x_k)$ or of the form $F(x_1,...,x_k) = y$. For each formula ϕ

of M, we shall construct a special formula ϕ_s of M such that ϕ is true in M iff ϕ_s is true in M for every assignment of meanings to the free variables. If will clearly suffice to do this for atomic ϕ. We use induction on the number n of occurrences of function symbols in ϕ. If $n = 0$, ϕ_s is ϕ. Otherwise, there is a formula ψ having n–1 occurrences of function symbols such that ϕ results from ψ by replacing the free occurrences of x by $F(x_1,...,x_k)$. We then take ϕ_s to be $\exists x(x = F(x_1,...,x_k) \mathbin{\&} \psi_s)$.

Now suppose that M is definable in M'. For each formula ϕ of M we shall construct a formula ϕ^* of M' such ϕ^* is true in M' iff ϕ is true in M when the free variables are assigned meanings in $|M|$. We can suppose that ϕ is special; otherwise we replace ϕ by ϕ_s. Let ψ be an atomic formula in ϕ not of the form $x = y$. If ψ is of the form $R(x_1,...,x_k)$, then R is definable by some χ in M' using $x_1,...,x_k$. Replace ϕ by χ. A similar procedure takes care of the case that ψ is $F(x_1,...,x_k) = y$. Finally, replace each part $\exists x\psi$ or $\forall x\psi$ of ϕ by $\exists x(\chi \mathbin{\&} \psi)$ or $\forall x(\chi \to \psi)$ where χ is a formula such that $|M|$ is defined by χ in M' using x. The resulting formula is ϕ^*. (The reader unfamiliar with logic is advised to check all the details.)

We are going to define a finte set Q of sentences true in M' with the following property: if N' is a structure for the language of M' in which all the sentences in Q are true, then there is a structure N for the language of M such that N is definable in N' using the same formulas used to define M in M'. It will follows that ϕ^* is true in N' iff ϕ is true in N. Let $|M|$ be defined by χ in M' using x. If F is a k–ary function symbol in the language of M, let the graph of F_M be defined by ψ_F in M' using $x_1,...,x_k,y$. The sentences of Q must insure that the set defined by χ in N' is non–empty and that the relation defined by ψ_F in N' is the graph of a function. Let $\chi(t)$ be obtained from χ by replacing the free occurrences of x by t; and let ψ_F' be obtained from ψ_F by replacing the free occurrences of y by y'. Then Q contains the sentence $\exists x\chi$, and, for each F, the two sentences

$$\forall x_1 \ldots \forall x_k(\chi(x_1) \,\&\, \ldots \,\&\, \chi(x_k) \rightarrow \exists y(\chi(y) \,\&\, \psi_F))$$

and $\forall x_1 \ldots \forall x_k \forall y \forall y'(\chi(x_1) \,\&\, \ldots \,\&\, \chi(x_k) \,\&\, \chi(y) \,\&\, \chi(y') \,\&\, \psi_F \,\&\, \psi'_F \rightarrow y = y')$.

11.3. PROPOSITION. If M is definable in M' and M is strongly undecidable, then M' is strongly undecidable.

Proof. Let M' be a model of T'; we must show that T' is undecidable. Since Q is finite and every sentence in Q is true in M', we may suppose by 11.1 that the sentences of Q are axioms of T'. Let T be the theory with the language of M whose axioms are all ϕ such that ϕ^* is a theorem of T'. For any such ϕ, ϕ^* is true in M'; so ϕ is true in M. Thus M is a model of T; so T is undecidable.

It is thus sufficient to show that a solution to the decision problem for T' would give a solution of the decision problem for T. We show this by showing that ϕ is a theorem of T iff ϕ^* is a theorem of T'. If ϕ^* is a theorem of T', then ϕ is an axiom of T and hence a theorem of T. Suppose that ϕ is a theorem of T; we must show that ϕ^* is true in every model N' of T'. Let N be definable in N' by the same formulas used to define M in M'. It is enough to show that ϕ is true in N; and for this, it is enough to show that N is a model of T. If ψ is an axiom of T, then ψ^* is a theorem of T'; so ψ^* is true in N'; so ψ is true in N. □

We now construct a theory PO. The language of PO consists of a binary relation symbol $<$. Then axioms of PO are

$$\forall x \neg(x < x)$$

and $$\forall x \forall y \forall z(x < y \,\&\, y < z \rightarrow x < z).$$

A model of PO is called a partially ordered set. (We have chosen to present partially ordered sets in terms of the $<$ relation; it would make no essential difference if we used the $\leq$ relation instead.)

We shall construct a strongly undecidable partially ordered set. Recall that we have constructed a strongly undecidable structure M whose language consists of a binary function symbol F. Hence by 11.2 and 11.3, it will suffice to construct a model M' of PO such that M is definable in an inessential extension of M'.

Let $M_1 = |M| \cup \{1,2,3\}$, where 1,2,3 are objects not in $|M|$. Let M_2 be the set of ordered pairs (x,i) where $x \in |M|$ and $i \in \{1,2,3\}$. Let M_3 be the set of ordered triples (x,y,z) such that $x,y,z \in |M|$ and $F_M(x,y) = z$. Let $|M'| = M_1 \cup M_2 \cup M_3$. We define $<_{M'}$ as follows. If $x \in M_1$, $w <_{M'} x$ is false for all w. If $<x,i> \in M_2$, then $w <_{M'} <x,i>$ holds for $w = x$ and $w = i$. If $<x,y,z> \in M_3$, then $w <_{M'} <x,y,z>$ if w is one of $<x,1>$, $<y,2>$, $<z,3>$, x, y, z, 1, 2, or 3. Clearly M' is a partially ordered set.

For $x,y,z \in |M'|$ we have

$$x \in |M| \leftrightarrow \forall y \neg(y < x) \;\&\; x \neq 1 \;\&\; x \neq 2 \;\&\; x \neq 3,$$

$$F(x,y) = z \leftrightarrow x,y,z \in |M| \;\&\; \exists u \exists x_1 \exists y_1 \exists z_1 (x < x_1 \;\&\; y < y_1 \;\&\; z < z_1 \;\&\; 1 < x_1 \;\&\; 2 < y_1 \;\&\; 3 < z_1 \;\&\; x_1 < u \;\&\; y_1 < u \;\&\; z_1 < u).$$

(In proving the second equivalence from right to left, one should first note that we must have $x_1, y_1, z_1 \in M_2$ and $u \in M_3$.) It follows easily that M is definable in M'', where M'' is an inessential expansion of M' formed by adding three new constants to represent 1, 2, and 3.

It follows that PO is undecidable. It also follows that a theory whose language consists of one binary relation symbol and which has no axioms is undecidable.

Many other strongly undecidable structures can be constructed by these methods. However, the proof that M is definable in M' often requires a very detailed analysis of M and M'.

12. Relative Recursion

Let Φ be a set of *total* functions. We generalize the notion of computable to allow us to use the values of the functions in Φ at any arguments we wish in the course of the computation. Following Turing, we picture the computation as taking place as follows. For each F in Φ, we are given an object called an oracle for F. During the computation, we may ask the oracle for $F(\vec{x})$ for any $\vec{x}$ which we have computed. The oracle supplies the value, which may then be used in the rest of the computation. A function which can be computed using oracles for the functions in Φ is said to be computable in Φ or relative to Φ. (Turing used this terminology because the oracle produces a value of the function without the use of an algorithm. However, one should not consider an oracle as a mystical object. We can think of it as an infinite set of file cards on each of which a value of the function is printed; we get the value at a particular set of arguments by searching through the file.)

We now extend the notion of recursive in a similar way. For each F in Φ, we introduce a new type of instruction, called an F–instruction. This instruction has the format

$$F(\mathfrak{X}i_1,\ldots,\mathfrak{X}i_k) \to \mathfrak{X}j$$

where $i_1,\ldots,i_k,j$ are distinct. If the machine executes this instruction when $x_1,\ldots,x_k$ are in $\mathfrak{X}i_1,\ldots,\mathfrak{X}i_k$ respectively, it replace the number in $\mathfrak{X}j$ by $F(x_1,\ldots,x_k)$ and increases the number in the counter by 1. The Φ–machine is obtained from the basic machine by adding all F–instructions for all F in Φ. We define the notion of a program computing a function for this machine as for the basic machine. A function is recursive in Φ or relative to Φ if it is computed by some program for the Φ–machine. Note that if Φ is empty, recursive in Φ is the same as recursive.

12.1. PROPOSITION. If $\Phi \subseteq \Psi$, then every function recursive in Φ is recursive in Ψ. □

12.2. PROPOSITION. Every function recursive in Φ is recursive in some finite subset of Φ.

Proof. Any program for the Φ–machine can contain F–instructions for only a finite number of F in Φ. □

We will now show that all of the results we have proved remain true if recursive is replaced by recursive in Φ, in some cases under the assumption that Φ is finite. In most cases, no change is required in the proofs. We shall run quickly through these results, except where the presence of Φ makes a significant difference.

First, the results of §4 may be extended from the basic machine to the Φ–machine. This leads to the following result.

12.3. TRANSITIVITY THEOREM. If every function in Ψ is recursive in Φ, then every function which is recursive in Ψ is recursive in Φ.

Proof. Let G be recursive in Ψ and let P be a program for the Ψ–machine which computes it. If P contains a F–instruction, then F is recursive in Φ; so we may replace the instruction by the macro for the Φ–machine with the same format. We thus obtain a program for the macro Φ–machine which computes G; so G is recursive in Φ by 4.2. □

12.4. COROLLARY. If every function in Φ is recursive in Ψ and every function in Ψ is recursive in Φ, then the same functions are recursive in Φ and in Ψ. □

The proof in §5 that the class of recursive functions is recursively closed extends to the relative case. We also have the following result.

12.5. PROPOSITION. Every function in Φ is recursive in Φ.

Proof. A function F in Φ is computed by the program consisting of the one instruction $F(\mathfrak{X}1, ..., \mathfrak{X}k) \to \mathfrak{X}0$. □

The extensions of the results of the next few sections depend only on the fact that the class of functions recursive in Φ is recursively closed and includes Φ. All of the results of §6 extend; i.e., all of the symbols proved recursive are recursive in Φ. In addition, names of functions and relations in Φ are recursive in Φ by 12.5. The results of §7 extend.

12.6. PROPOSITION. If Ψ consists of the contractions of the functions in Φ, then the same functions are recursive in Φ and in Ψ.

Proof. By 12.4 and the contraction formulas. □

In §8 we run into a problem; if Φ is too large, we cannot assign codes to all of the F–instructions for the Φ–machine. We shall therefore now suppose that Φ is finite. By 12.6, we may suppose that Φ is a finite sequence $H_1, ..., H_m$ of reals. To the H_r–instruction $H_r(\mathfrak{X}i) \to \mathfrak{X}j$ we assign the code $<3,i,j,r>$. Except for this, codes are defined as before.

We can now extend §8 straightforwardly. Since the functions and relations defined there now depend on Φ, we add a superscript Φ to their names. Thus $T_k^{\Phi}(e,\vec{x},y)$ means that e is the code of a program P for the Φ–machine, and y is the code of the P–computation from $\vec{x}$. Then T_k^{Φ} is recursive in Φ. Note, however, that we do not need a U^{Φ}, since the old U still serves.

The Normal Form Theorem now becomes $\{e\}^{\Phi}(\vec{x}) \simeq U(\mu y T_k^{\Phi}(e,\vec{x},y))$. We say that e is a <u>Φ–index</u> of F if $F(\vec{x}) \simeq \{e\}^{\Phi}(\vec{x})$ for all $\vec{x}$. Then a function has a Φ–index iff it is recursive in Φ.

Results which mention indices can be relativized to Φ only for finite Φ, since it is only for finite Φ that Φ–index is defined. Sometimes a result may not mention indices, but its proof may use indices. In this case, all we can immediately say is that the result holds when relativized to a finite Φ. In some cases we can then extend the result to all Φ by using 12.2.

An example where such an extension is possible is 8.3. Our extension says that the class of functions recursive in Φ is the smallest recursively closed class which includes Φ. We can immediately say this is true for finite Φ. Now let Φ be arbitrary and let Ψ be a be a recursively closed class including Φ. We want to show that if F is recursive in Φ, then F is in Ψ. Now F is recursive in a finite subset of Φ. Since Ψ includes this finite subset, F is in Ψ.

We now consider how T_k^{Φ} depends on Φ. When the definitions of §8 are relativized to Φ, the H_r appear explicitly only in the definition of *Reg*. The definition of this function has a new clause for each r:

$$Reg(j,e,x,n+1) = H_r(Reg((i)_1,e,x,n)) \text{ if } (i)_0 = 3 \,\&\, (i)_3 = r \,\&\, (i)_2 = j.$$

This means that in the definition of $T_k^{\Phi}(e,\vec{x},y)$, H_r appears only in contexts $H_r(X)$ where X designates a number appearing in a register during the P–computation from $\vec{x}$ and hence $< y$. Thus we may replace $H_r(X)$ by $(\overline{H_r}(y))_X$.

If Φ is $H_1,\ldots,H_m$, we write $\overline{\Phi}(z)$ for $\overline{H_1}(z),\ldots,\overline{H_m}(z)$. The above can be summarized as follows: there is a recursive relation $T_{k,m}$ such that

$$(1) \qquad T_k^{\Phi}(e,\vec{x},y) \leftrightarrow T_{k,m}(e,\vec{x},y,\overline{\Phi}(y)).$$

Thus if $\{e\}^{\Phi}(\vec{x}) \simeq z$ with computation number y, and $\overline{\Phi}(y) = \overline{\Phi'}(y)$, then $\{e\}^{\Phi'}(\vec{x}) \simeq z$.

13. The Arithmetical Hierarchy

We are now going to study the effect of using unbounded quantifiers in definitions of relations. From now on, we agree that n designates a non–zero number. The results of this section are due to Kleene.

A relation R is arithmetical if it has an explicit definition

$$(1) \qquad R(\vec{x}) \leftrightarrow Qy_1 \ldots Qy_n P(\vec{x},y_1,\ldots,y_n)$$

where each Qy_i is either $\exists y_i$ or $\forall y_i$ and P is recursive. We call $Qy_1\ldots Qy_n$ the prefix and $P(\vec{x},y_1,\ldots,y_n)$ the matrix of the definition. We are chiefly interested in the prefix, since it measures how far the definition is from being recursive.

We shall first see how prefixes can be simplified. As z runs through all number, $(z)_0,(z)_1$ runs through all pairs of numbers. It follows that

$$\forall x \forall y R(x,y) \leftrightarrow \forall z R((z)_0,(z)_1)$$

and

$$\exists x \exists y R(x,y) \leftrightarrow \exists z R((z)_0,(z)_1).$$

Using these equivalences, we can replace two adjacent universal quantifiers in a prefix by a single such quantifier, and similarly for existential quantifiers. For example, a definition

$$R(x) \leftrightarrow \forall y \forall z \exists v P(x,y,z,v)$$

can be replaced by

$$R(x) \leftrightarrow \forall w \exists v P(x,(w)_0,(w)_1,v).$$

Of course, the matrix has changed; but it is still a recursive function of its variables because $(w)_0$ and $(w)_1$ are recursive functions of w. This sort of simplification of a prefix is called contraction of quantifiers.

A prefix is alternating if it does not contain two successive existential quantifiers or two successive universal quantifiers. A prefix is Π_n^0 if it is alternating, has n quantifiers, and begins with $\forall$. A prefix is Σ_n^0 if it is alternating, has n quantifiers, and begins with $\exists$. A relation is Π_n^0 if it has an explicit definition with a Π_n^0 prefix and a recursive matrix; similarly for Σ_n^0. A relation is Δ_n^0 if it is both Π_n^0 and Σ_n^0. We sometimes use Π_n^0 for the class of Π_n^0 relations; similarly for Σ_n^0 and Δ_n^0.

13.1. Proposition. Every arithmetical relation is Π_n^0 or Σ_n^0 for some n.

Proof. By contraction of quantifiers. □

13.2. Proposition. If R is Π_n^0 or Σ_n^0, then R is Δ_k^0 for every $k > n$. If R is recursive, then R is Δ_n^0 for all n.

Proof. By adding superfluous quantifiers. For example, suppose that R is Π_2^0; say $R(x) \leftrightarrow \forall y \exists z P(x,y,z)$. To show that R is Δ_3^0, we note that

$$R(x) \leftrightarrow \forall y \exists z \forall w P(x,y,z)$$

$$\leftrightarrow \exists w \forall y \exists z P(x,y,z). \ \square$$

A relation P is many-one reducible, or simply reducible, to a relation Q if it has a definition

$$P(\vec{x}) \leftrightarrow Q(F_1(\vec{x}),\dots,F_n(\vec{x}))$$

where each F_i is total and recursive. If P is reducible to Q and Q is recursive, then P is recursive. From this we obtain the following result.

13.3. Proposition. If P is reducible to Q and Q is Π_n^0, then P is Π_n^0. The same holds with Σ_n^0 or Δ_n^0 in place of Π_n^0. □

The contraction formulas show that R and $<R>$ are reducible to one another; so R is Π^0_n iff $<R>$ is Π^0_n, and similarly for Σ^0_n and Δ^0_n.

We now consider the effect of applying propositional connectives to arithmetical relations. The key tools are the <u>prenex rules</u>, which are certain rules for bringing quantifiers to the front of an expression. They are

$$\neg QxR(x) \leftrightarrow Q'x\neg R(x),$$
$$QxR(x) \vee P \leftrightarrow Qx(R(x) \vee P),$$
$$P \vee QxR(x) \leftrightarrow Qx(P \vee R(x)),$$
$$QxR(x) \ \&\ P \leftrightarrow Qx(R(x) \ \&\ P),$$
$$P \ \&\ QxR(x) \leftrightarrow Qx(P \ \&\ R(x)),$$

where Q is either $\forall$ or $\exists$ and Q' is $\exists$ if Q is $\forall$ and $\forall$ if Q is $\exists$. These rules are well known and easily seen to be valid.

From the first rule (and the fact that $\neg$ is a recursive symbol) we see that the negation of a Π^0_n relation is Σ^0_n and the negation of a Σ^0_n relation is Π^0_n. For example, to see that the negation of a Π^0_2 relation is Σ^0_2, note that by the prenex rules

$$\neg\forall x\exists yP \leftrightarrow \exists x\forall y\neg P.$$

The next four rules together with contraction of quantifiers show that the disjunction and conjunction of two Π^0_n relations is Π^0_n; and similarly for Σ^0_n. For example, to treat the disjunction of two Π^0_2 relations, observe that by the prenex rules

$$\forall x\exists yP \vee \forall z\exists wQ \leftrightarrow \forall x\forall z\exists y\exists w(P \vee Q)$$

and then use contraction of quantifiers.

Now consider a definition $R(\vec{x}) \leftrightarrow \forall yP(\vec{x},y)$ where P is arithmetical. By replacing P in this definition by the right side of the definition of P and then using contraction of quantifiers if possible, we see that if P is Π^0_n, then R is Π^0_n; and if P is Σ^0_n, then R is Π^0_{n+1}. A similar result holds if $\forall y$ is replaced by $\exists y$.

Now we consider the effect of bounded quantifiers. We need the following equivalences:

$$(\forall y < x)\forall z R(x,y,z) \leftrightarrow \forall z(\forall y < x)R(x,y,z),$$

$$(\exists y < x)\exists z R(x,y,z) \leftrightarrow \exists z(\exists y < x)R(x,y,z),$$

$$(\forall y < x)\exists z R(x,y,z) \leftrightarrow \exists z(\forall y < x)R(x,y,(z)_y),$$

$$(\exists y < x)\forall z R(x,y,z) \leftrightarrow \forall z(\exists y < x)R(x,y,(z)_y).$$

The first two of these are obvious. Both sides of the third says that there is a sequence $z_0, \ldots, z_{x-1}$ such that $R(x,y,z_y)$ for all $y < x$. Now replace R by $\neg R$ in the third equivalence, bring the negation signs to the front by means of the prenex rules, and then drop the negations signs from the front of both sides of the equivalence. We then obtain the fourth equivalence.

Now consider a definition $R(\vec{x},z) \leftrightarrow (Qy < z)P(\vec{x},y,z)$ where P is arithmetical. Substitute the right side of the definition of P for P. We can then apply the above equivalences to bring all of the unbounded quantifiers to the left of $(Qy < z)$. Since bounded quantifiers are recursive, we may now consider $(Qy < z)$ as part of the matrix. It follows that if P is Π^0_n, then so is R; and similarly for Σ^0_n.

We can summarize our results in the following table, which gives the classification of various combinations of P and Q in terms of the classifications of P and Q.

P,Q	$\neg P$	$P \vee Q$	$P \& Q$	$\forall x P$	$\exists x P$	$(Qx<y)P$
Π^0_n	Σ^0_n	Π^0_n	Π^0_n	Π^0_n	Σ^0_{n+1}	Π^0_n
Σ^0_n	Π^0_n	Σ^0_n	Σ^0_n	Π^0_{n+1}	Σ^0_n	Σ^0_n
Δ^0_n	Δ^0_n	Δ^0_n	Δ^0_n	Π^0_n	Σ^0_n	Δ^0_n

(The last row of the table follows from the first two rows.) To treat the case in which P and Q do not have the same classification, we use 13.2. For example, if P is Π^0_2 and Q is Σ^0_2, then P and Q are Δ^0_3, and we can use the last row of the table. To treat $\rightarrow$ and $\leftrightarrow$, we replace $X \rightarrow Y$ by $\neg X \vee Y$ and $X \leftrightarrow Y$ by $(X \rightarrow Y)$ & $(Y \rightarrow X)$. Every recursion theorist should learn this table.

The classification of the arithmetical relations into Π^0_n and Σ^0_n relations is called the arithmetical hierarchy. We have not yet shown that the classes in this hierarchy are distinct.

Let Φ be a class of k–ary relations. We say that a $(k+1)$–ary relation Q enumerates Φ if for every R in Φ, there is a e such that $R(\vec{x}) \leftrightarrow Q(\vec{x},e)$ for all $\vec{x}$.

13.4. Arithmetical Enumeration Theorem. For every n and k, there is a $(k+1)$–ary Π^0_n relation which enumerates the class of k–ary Π^0_n relations; and similarly with Σ^0_n for Π^0_n.

Proof. We suppose that $n = 2$; other values of n are similar. Suppose that R is Π^0_2; say $R(\vec{x}) \leftrightarrow \forall y \exists z P(\vec{x},y,z)$ where P is recursive. Let e be an index of χ_P. Then

$$R(\vec{x}) \leftrightarrow \forall y \exists z(\{e\}(\vec{x},y,z) \simeq 0)$$
$$\leftrightarrow \forall y \exists z \exists s(\{e\}_s(\vec{x},y,z) \simeq 0).$$

If we let $Q(\vec{x},e)$ be the right side of this equation, then Q is Π^0_2 by 8.4 and the table; so Q is the desired enumerating relation for Π^0_2. By the table, $\neg Q$ is the desired enumerating relation for Σ^0_2. □

Suppose that R is a binary relation which enumerates the class Φ of sets. We can use the diagonal method to define a set A which is not in Φ. Since we want $A(x)$ to be different from $R(x,e)$ when $x = e$, we set $A(e) \leftrightarrow \neg R(e,e)$. To put it another way, let D be the diagonal set defined by $D(e) \leftrightarrow R(e,e)$. Then if R enumerates Φ, $\neg D$ is not in Φ.

13.5. Arithmetical Hierarchy Theorem. For each n, there is a Π^0_n unary relation which is not Σ^0_n, hence not Π^0_k or Σ^0_k for any $k < n$. The same holds with Π^0_n and Σ^0_n interchanged.

Proof. We prove the first half; the second half is similar. Let P be a binary Π^0_n relation which enumerates the class of unary Π^0_n relations, and define $D(e) \leftrightarrow P(e,e)$. By 13.3, D is Π^0_n. By the above discussion, $\neg D$ is not Π^0_n; so by the table, D is not Σ^0_n. By 13.2, D is not Π^0_k or Σ^0_k for any $k < n$. □

The Arithmetical Hierarchy Theorem shows that there are no inclusions among the classes Π_n^0 and Σ_n^0 other than those given by 13.2.

The Arithmetical Enumeration Theorem is false for Δ_n^0 relations; for if it were true, we could use the proof of the Arithmetical Hierarchy Theorem to show that there is a Δ_n^0 relation which is not Δ_n^0.

Let Φ be a set of total functions. If Q is any concept defined in terms of recursive functions, we can obtain a definition of Q in Φ or relative to Φ by replacing recursive everywhere in the definition of Q by recursive in Φ. For example, R is arithmetical in Φ if it has a definition (1) where P is recursive in Φ; and R is Π_n^0 in Φ if it has such a definition in which the prefix is Π_n^0. We shall assume that this is done for all past and future definitions.

Now let us consider how the results of this section extend to the relativized case. Up to the Enumeration Theorem, everything extends without problems. The rest extends to finite Φ but not to arbitrary Φ. For example, if Φ is the set of all reals, then every unary relation is recursive in Φ and hence Π_n^0 and Σ_n^0 in Φ for all n. Thus the Hierarchy Theorem fails. Since the Hierarchy Theorem is a consequence of the Enumeration Theorem, the Enumeration Theorem also fails.

14. Recursively Enumerable Relations

A relation R is semicomputable if there is an algorithm which, when applied to the inputs $\vec{x}$, gives an output iff $R(\vec{x})$. If F is the function computed by the algorithm, then the algorithm applied to $\vec{x}$ gives an output iff $\vec{x}$ is in the domain of F. Hence R is semicomputable iff it is the domain of a computable function.

As an example, let A be the set of n such that $x^n + y^n = z^n$ holds for some positive integers x, y, and z. Then A is semicomputable; the algorithm with input n tests each triple (x,y,z) in turn to see if $x^n + y^n = z^n$. On the other hand, it is not known if A is computable.

A relation is recursively enumerable (abbreviated RE) if it is the domain of a recursive function. By the above and Church's Thesis, a relation is RE iff it is semicomputable.

We let W_e be the domain of the function $\{e\}$. We say that e is an index of the relation R if R is W_e. (Note that this is not the same as being an index of the function χ_R.) Clearly a relation has an index iff it is RE. By the Normal Form Theorem, we have

$$(1) \qquad W_e(\vec{x}) \leftrightarrow \exists y T_k(e,\vec{x},y).$$

14.1. **Proposition.** A relation is RE iff it is Σ^0_1.

Proof. If R is RE, it is W_e for some e and hence Σ^0_1 by (1). Suppose that R is Σ^0_1; say $R(\vec{x}) \leftrightarrow \exists y P(\vec{x},y)$ with P recursive. Then R is the domain of the recursive function F defined by $F(\vec{x}) \simeq \mu y P(\vec{x},y)$ and hence is RE. □

We often use 14.1 tacitly. In particular, we use it to apply the results of the last section to RE relations. By the Enumeration Theorem, there is a $(k+1)$–ary RE relation R which enumerates the class of k–ary RE relations. In fact, we can define such an R by $R(\vec{x},e) \leftrightarrow W_e(\vec{x})$; this is RE by (1).

By the Arithmetical Hierarchy Theorem, there is an RE set which is not recursive. In fact, the proof of that theorem shows that such a set D is defined by $D(e) \leftrightarrow W_e(e)$.

We let $W_{e,s}$ be the domain of $\{e\}_s$. Then $W_e(\vec{x})$ iff $W_{e,s}(\vec{x})$ for some s; in this case, $W_{e,s}(\vec{x})$ for all $s \geq y$, where y is the computation number of $\{e\}(\vec{x})$. By 8.4, $W_{e,s}(\vec{x})$ is a recursive relation of e,s, and $\vec{x}$. Note also that if $W_{e,s}(\vec{x})$, then each x_i is less than the computation number of $\{e\}(\vec{x})$ and hence less than s. Thus $W_{e,s}$ is finite.

14.2. **RE Parameter Theorem.** If R is a $(k+m)$–ary RE relation, there is a recursive total function S such that

$$W_{S(y_1,\dots,y_m)}(\vec{x}) \leftrightarrow R(\vec{x},y_1,\dots,y_m)$$

for all $\vec{x},y_1,\dots,y_m$.

Proof. This is easily proved from the Parameter Theorem. □

We can use implicit definitions to define RE relations. Thus suppose that we want to find an RE relation R with an index e such that $R(\vec{x}) \leftrightarrow P(\vec{x},e)$, where P is RE. Let P be the domain of the recursive function G. By the Recursion Theorem, we can find a recursive F with an index e such that $F(\vec{x}) \simeq G(\vec{x},e)$. We then take R to be the domain of F.

A selector for a $(k+1)$–ary relation R is a k–ary function F such that for each $\vec{x}$, $F(\vec{x})$ is defined iff $\exists y R(\vec{x},y)$; and, in this case, $F(\vec{x})$ is a y such that $R(\vec{x},y)$.

14.3. Selector Theorem. Every $(k+1)$–ary RE relation has a recursive selector.

Proof. Let $R(\vec{x},y) \leftrightarrow \exists z P(\vec{x},y,z)$ with P recursive. Then a recursive selector F for R is defined by

$$F(\vec{x}) \simeq (\mu w P(\vec{x},(w)_0,(w)_1))_0. \ \square$$

If F is k–ary, the graph of F, designated by $\mathcal{G}_F$, is the $(k+1)$–ary relation defined by

$$\mathcal{G}_F(\vec{x},y) \leftrightarrow F(\vec{x}) \simeq y.$$

The next theorem shows how to characterize recursive functions in terms of recursive relations.

14.4. Graph Theorem. A function F is recursive iff its graph is RE. A total function F is recursive iff its graph is recursive.

Proof. Let F be recursive and let e be an index of F. Then

$$\begin{aligned}\mathcal{G}_F(\vec{x},y) &\leftrightarrow \{e\}(\vec{x}) \simeq y\\ &\leftrightarrow \exists s(\{e\}_s(\vec{x}) \simeq y).\end{aligned}$$

Thus $\mathcal{G}_F$ is RE by 8.4. If F is total, then the definition

$$\mathcal{G}_F(\vec{x},y) \leftrightarrow F(\vec{x}) = y$$

shows that $\mathcal{G}_F$ is recursive. If $\mathcal{G}_F$ is RE, then it has a recursive selector. But the only selector for $\mathcal{G}_F$ is F. □

As an application, we prove a more general result on definition by cases of recursive functions.

14.5. PROPOSITION. Let $R_1, \ldots, R_n$ be RE relations such that for every $\vec{x}$, at most one of $R_1(\vec{x}), \ldots, R_m(\vec{x})$ is true. Let $F_1, \ldots, F_n$ be recursive functions, and define F by

$$\begin{aligned} F(\vec{x}) &\simeq F_1(\vec{x}) \quad \text{if } R_1(\vec{x}), \\ &\simeq \ldots \\ &\simeq F_n(\vec{x}) \quad \text{if } R_n(\vec{x}), \end{aligned}$$

where it is understood that $F(\vec{x})$ is undefined if none of $R_1(\vec{x}), \ldots, R_n(\vec{x})$ is true. Then F is recursive.

Proof. We have

$$\mathcal{G}_F(\vec{x},y) \leftrightarrow (\mathcal{G}_{F_1}(\vec{x},y) \;\&\; R_1(\vec{x})) \vee \ldots \vee (\mathcal{G}_{F_n}(\vec{x},y) \;\&\; R_n(\vec{x})).$$

By the Graph Theorem and the table, the right side is RE; so F is recursive by the Graph Theorem. □

The next result characterizes recursive relations in terms of RE relations.

14.6. PROPOSITION. A relation R is recursive iff both R and $\neg R$ are RE.

Proof. If R is recursive, then $\neg R$ is recursive; so R and $\neg R$ are RE by 13.2. Now

$$\mathcal{G}_{\chi_R}(\vec{x},y) \leftrightarrow (R(\vec{x}) \;\&\; y = 0) \vee (\neg R(\vec{x}) \;\&\; y = 1).$$

If R and $\neg R$ are RE, this equivalence and the table show that $\mathcal{G}_{\chi_R}$ is RE; so R is recursive by the Graph Theorem. □

14.7. PROPOSITION. A non-empty set A is RE iff it is the range of a recursive real. An infinite set A is RE iff it is the range of a one-one recursive real.

Proof. If F is a recursive real, its range A is defined by $y \in A \leftrightarrow$

$\exists x(F(x) = y)$; so A is RE. Now let A be an RE set. Let e be an index of A and let $a \in A$. Define a recursive real F by

$$F(x) \simeq (x)_0 \quad \text{if } T_1(e,(x)_0,(x)_1),$$
$$\simeq a \quad \text{otherwise.}$$

Clearly the range of F is A. Now suppose that A is also infinite. Define $G(n) = F(H(n))$ where $H(n)$ is the least x such that $F(x) \neq F(H(m))$ for all $m < n$. Then H is defined by course-of-values recursion using only recursive symbols and hence is recursive; so G is recursive. Clearly G is one–one and has range A. □

All of the results of this section relativize to any finite Φ. For Φ a finite sequence of reals, we let W_e^{Φ} be the domain of $\{e\}^{\Phi}$, and say that e is a <u>Φ–index</u> of W_e^{Φ}. Then (1) becomes

$$(2) \qquad W_e^{\Phi}(\vec{x}) \leftrightarrow \exists y T_k^{\Phi}(e,\vec{x},y).$$

We can use (1) of §12 to rewrite this as

$$(3) \qquad W_e^{\Phi}(\vec{x}) \leftrightarrow \exists y T_{k,m}(e,\vec{x},y,\bar{\Phi}(y)).$$

Using 12.2, we see that a relation is RE in Φ iff it is RE in a finite subset of Φ; and similarly for Σ_1^0. It follows that 14.1 relativizes to arbitrary Φ. Similarly, 14.3 through 14.6 relativize to arbitrary Φ.

14.8. Proposition (Post). A relation is Σ_{n+1}^0 iff it is RE in Π_n^0.

Proof. If R is Σ_{n+1}^0, then $R(\vec{x}) \leftrightarrow \exists y P(\vec{x},y)$ where P is Π_n^0. Then R is RE in P and hence in Π_n^0.

Now suppose that R is RE in Π_n^0. By 12.2, R is RE in a finite subset of Π_n^0. By the remark after 13.3 and 12.6, we may suppose the relations in Φ are unary. To simplify the notation, suppose that Φ consists of one relation P. By (3), we have for some recursive Q

$$R(\vec{x}) \leftrightarrow \exists y Q(\overline{\chi_P}(y),\vec{x},y)$$
$$\leftrightarrow \exists y \exists z(z = \overline{\chi_P}(y) \ \& \ Q(z,\vec{x},y)).$$

If we can show that $z = \overline{\chi_P}(y)$ is Σ_{n+1}^0, it will follow by the table that R is Σ_{n+1}^0. Now

$$z = \overline{\chi_P}(y) \leftrightarrow Seq(z) \,\&\, lh(z) = y \,\&\, (\forall i < y)((z)_i = \chi_P(i)).$$

Hence by the table, it will suffice to show that $w = \chi_P(i)$ is Σ^0_{n+1}. Since P is Π^0_n, this follows from

$$w = \chi_P(i) \leftrightarrow (w = 1 \,\&\, P(i)) \vee (w = 0 \,\&\, \neg P(i))$$

and the table. □

14.9. COROLLARY. A relation is Δ^0_{n+1} iff it is recursive in Π^0_n.

Proof. A relation R is Δ^0_{n+1} iff both R and $\neg R$ are Σ^0_{n+1}; hence, by Post's Theorem, iff both R and $\neg R$ are RE in Π^0_n. By the relativized version of 14.6, this holds iff R is recursive in Π^0_n. □

Since $\neg R$ is recursive in R and $R = \neg\neg R$ is recursive in $\neg R$, 12.4 and the table show that we can replace Π^0_n by Σ^0_n in both Post's Theorem and its corollary.

15. Degrees

If F and G are total functions, we let $F \leq_R G$ mean that F is recursive in G. By 12.5,

$$(1) \qquad F \leq_R F;$$

and by the Transitivity Theorem

$$(2) \qquad F \leq_R G \,\&\, G \leq_R H \rightarrow F \leq_R H.$$

Let $F \equiv_R G$ mean $F \leq_R G \,\&\, G \leq_R F$. It follows from (1) and (2) that $\equiv_R$ is an equivalence relation. The equivalence class of F is called the degree of F and is designated by dg F. By a degree, we mean the degree of some total function. We use small boldface letters, usually **a**, **b**, **c**, and **d**, for degrees.

We let $\text{dg}(F) \leq \text{dg}(G)$ if $F \leq_R G$. By (2), this depends only on dg(F) and dg(G), not on the choice of F and G in these equivalence classes. It follows from (1) and (2) that $\leq$ is a partial ordering of the degrees, i.e., that

$$\mathbf{a} \leq \mathbf{a},$$

$$\mathbf{a} \leq \mathbf{b} \,\&\, \mathbf{b} \leq \mathbf{a} \rightarrow \mathbf{a} = \mathbf{b},$$

$$\mathbf{a} \leq \mathbf{b} \mathbin{\&} \mathbf{b} \leq \mathbf{c} \rightarrow \mathbf{a} \leq \mathbf{c}.$$

The degree of F is, roughly speaking, a measure of the difficulty of computing F; and dg $F \leq$ dg G means that F is at least as easy to compute as G.

By the degree of a relation R, we mean the degree of χ_R; we will sometimes say (with an abuse of language) that R belongs to that degree. Every degree is the degree of a relation; for a total function has the same degree as its graph (because of the equivalences $\mathcal{G}_F(\vec{x},y) \leftrightarrow F(\vec{x}) = y$ and $F(\vec{x}) = \mu y \, \mathcal{G}_F(\vec{x},y)$.) A total function or relation has the same degree as its contraction by the contraction equations; so every degree contains a real and a set.

Let $\mathbf{0}$ be the class of recursive total functions. It is easy to see that $\mathbf{0}$ is a degree and that $\mathbf{0} \leq \mathbf{a}$ for every degree $\mathbf{a}$. Thus $\mathbf{0}$ is the smallest degree.

15.1. **Proposition.** Every finite class of degrees has a least upper bound.

Proof. Let the degrees in the class be dg(F_1), ... dg(F_k) where F_1, ..., F_k are reals. Set $G(x) = \langle F_1(x),\dots,F_k(x)\rangle$. Since $F_i(x) = (G(x))_{i-1}$, F_i is recursive in G; so dg G is an upper bound of the set. Now let dg(H) be any upper bound. Then each F_i is recursive in H. Since G is recursive in $F_1,\dots,F_k$, it is recursive in H; so dg $G \leq$ dg H, as required. □

Most of the rest of degree theory consists of showing that the partial ordering of the degrees fails to have some nice properties. We shall illustrate the idea by proving that it is not a linear ordering.

15.2. **Proposition** (**Kleene–Post**). There are degrees $\mathbf{a}$ and $\mathbf{b}$ such that neither $\mathbf{a} \leq \mathbf{b}$ nor $\mathbf{b} \leq \mathbf{a}$.

Proof. It suffices to produce reals F and G such that neither $F \leq_R G$ nor $G \leq_R F$. We break this down into infinitely many conditions which we wish to satisfy. Condition C_{2e} is that $F \neq \{e\}^G$; condition C_{2e+1} is that $G \neq \{e\}^F$.

We define F and G in infinitely many steps, at each of which we define finitely many values of F and G. At step e we ensure that C_e is satisfied. We shall only describe step $2e$, since step $2e+1$ is similar.

Suppose we are at the beginning of step $2e$. Let x be the least number such that $F(x)$ is not yet defined. Suppose first that there is a z in Seq such that $T_{1,1}(e,x,z)$ and such that $(z)_i = G(i)$ for every $i < lh(z)$ such that $G(i)$ is defined. Define $G(i) = (z)_i$ for every $i < lh(z)$ such that $G(i)$ is not yet defined. Then $\overline{G}(lh(z)) = z$ and hence $T_{1,1}(e,x,\overline{G}(lh(z)))$. It follows that $\{e\}^G(x) = U(lh(z))$. We set $F(x) = 1 \dot{-} U(lh(z))$, so that $F(x) \neq \{e\}^G(x)$.

Now suppose that there is no such z. Then we know that $\{e\}^G(x)$ will be undefined, so that we will have $F \neq \{e\}^G$. We set $F(x) = 0$ and define no new values of G.

It remains to show that F and G are total. At step $2e$, we defined F at the smallest argument for which it was not already defined. It follows that F is total. Similarly, the action at step $2e+1$ makes G total. □

A degree **a** is <u>RE</u> if it contains an RE relation. Since the contraction of an RE relation is an RE set, every RE degree contains an RE set. (However, every RE degree other than **0** contains sets which are not RE.)

The degree of $W_e(x)$, considered as a relation of e and x, is designated by **0'**. This degree is RE; and, since every RE set is clearly recursive in this relation, it is the largest RE degree. Since there is an RE set which is not recursive, there is an RE degree other than **0**; so **0** < **0'**.

We shall show that there is a connection between **0'** and limits. An infinite sequence $\{F_s\}$ of reals is <u>recursive</u> if the function G defined by $G(s,x) \simeq F_s(x)$ is recursive. We say F is the <u>limit</u> of $\{F_s\}$ if for each x, there is an s_0 such that $F_s(x) = F(x)$ for all $s \geq s_0$.

15.3. LEMMA. *If* $\{F_s\}$ *is a recursive sequence of reals, then* $\{e\}_s^{F_s}(\vec{x}) \simeq z$ *and* $\{e\}_s^{F_s}(x)$ *is defined are recursive relations of* $e,s,\vec{x},z$. *If* F *is the limit of* $\{F_s\}$ *and* $\{e\}^F(x) \simeq z$, *then* $\{e\}_s^{F_s}(x) \simeq z$ *for all sufficiently large* s.

Proof. Making use of (1) of §12,

$$\{e\}_s^{F_s}(\vec{x}) \simeq z \leftrightarrow (\exists y < s)(T_{k,m}(e,\vec{x},y,\overline{F}_s(y)) \;\&\; U(y) \simeq z),$$

$$\{e\}_s^{F_s}(\vec{x}) \textit{ is defined} \leftrightarrow (\exists y < s)T_{k,m}(e,\vec{x},y,\overline{F}_s(y)).$$

If $G(s,x) = F_s(x)$, $\overline{G}(s,x) = \overline{F}_s(x)$; so $\overline{F}_s(x)$ is a recursive function of s and x. This proves the first part of the proposition. Suppose that $\{e\}^F(\vec{x}) \simeq z$ and that y is the computation number of $\{e\}^F(\vec{x})$. Again using (1) of §12, $\{e\}^{F_s}(\vec{x}) \simeq z$ if $\overline{F}_s(y) = \overline{F}(y)$, which is true for large s. But then $\{e\}_s^{F_s}(x) \simeq z$ for $s > y$. □

15.4. Limit Lemma. A real is the limit of a recursive sequence of reals iff it has degree $\leq \mathbf{0}'$.

Proof. Let F be the limit of the recursive sequence $\{F_s\}$, and define an RE R by

$$R(n,x) \leftrightarrow \exists m(m > n \;\&\; F_m(x) \neq F_n(x)).$$

Setting $H(x) \simeq \mu n \neg R(n,x)$, H is total and $F(x) \simeq F_{H(n)}(x)$. It follows that F is recursive in R; so dg $F \leq$ dg $R \leq \mathbf{0}'$.

Now suppose that dg $F \leq \mathbf{0}'$. Then F is recursive in an RE set. Say that $F = \{f\}^{W_e}$. Define $F_s(x) = \{f\}_s^{W_{e,s}}(x)$ if the right side is defined and $F_s(x) = 0$ otherwise. Since $\{W_{e,s}\}$ is recursive and has limit W_e, 15.3 shows that $\{F_s\}$ is recursive and has limit F. □

The problem of whether there is an RE degree other than $\mathbf{0}$ and $\mathbf{0}'$ is known as Post's Problem. We solve this problem by extending 15.2 to RE degrees. The method used is called the priority method.

By an RE construction, we mean a construction in which at each step, a finite number of numbers are put into a set A, and the construction is recursive (i.e., x is put into A at step s is a recursive function of x and s). Then A is RE, since

$$x \in A \leftrightarrow \exists s(x \text{ is put into } A \text{ at step } s).$$

15.5. PROPOSITION (FRIEDBERG–MUCHNIK). There are RE degrees **a** and **b** such that neither $\mathbf{a} \leq \mathbf{b}$ nor $\mathbf{b} \leq \mathbf{a}$.

Proof. We shall use an RE construction to construct RE sets A and B such that neither $A \leq_R B$ nor $B \leq_R A$. We shall use Church's Thesis to verify that the construction is recursive; this is clearly a non–essential use of Church's Thesis. We wish to satisfy the conditions C_e, where C_{2e} is $A \neq \{e\}^B$ and C_{2e+1} is $B \neq \{e\}^A$. We shall discuss only C_{2e}; C_{2e+1} is treated similarly but with A and B interchanged.

Let us consider first how we could make C_{2e} hold if we had no other conditions to worry about. Pick an x; we will make C_{2e} hold by insuring that $x \in A$ iff $\{e\}^B(x) \simeq 1$. Let B^s be the (finite) set of number put into B before step s. At step s, if x has not yet been put in A, we see if $\{e\}_s^{B^s}(x) \simeq 1$; we can do this effectively by 15.3. If $\{e\}_s^{B^s}(x) \simeq 1$, we put x into A, and agree not to put any numbers $< s$ into B at step s or later; otherwise, we do nothing at step s. If $\{e\}^B(x) \simeq 1$, then $\{e\}_s^{B^s}(x) \simeq 1$ for all sufficiently large s by 15.3; so $x \in A$. Suppose that $x \in A$; say x is put into A at step s. Then $\{e\}^{B^s}(x) \simeq 1$ with computation number $z < s$. Since no number $< s$ is put into B at step s or later, $\overline{\chi}_{B^s}(z) = \overline{\chi}_B(z)$; so $\{e\}^B(x) \simeq 1$.

When we try to treat all of the conditions at once, we run into conflicts; it may happen that C_e wants to put a number x into A and C_f wants to keep x out of A. We resolve this conflict be giving priority to the lower numbered condition. Thus if $e < f$, we put x into A; if $f < e$, we keep x out of A. (No condition will conflict with itself.)

Let $r_s(e)$ be the largest $t < s$ such that a number is put into A or B by C_e at step t (or 0 if there is no such t). Let $R_s(e)$ be the maximum of the $r_s(f)$ for $f < e$, and let $x_s(e) = \langle e, R_s(e)\rangle$. Then $x_s(e)$ is known at the beginning of step s.

(The numbers $< r_s(e)$ are the numbers C_e wants to keep out of A and B at step s. Hence C_e can put $x_s(e) > R_s(e)$ into A or B at s without violating our priorities.)

Now we describe step s. Let $f = (s)_0$ and $x = x_s(f)$. If $f = 2e$ and $\{e\}_s^{B^s}(x) \simeq 1$, we put x in A (unless it has been put in A earlier); if $f = 2e+1$ and $\{e\}_s^{A^s}(x) \simeq 1$, we put x in B.

We now have to prove that the construction works, i.e., that all of the conditions are satisfied. We shall first prove that for each e, there are only finitely many numbers $x_s(e)$. The proof is by induction on e. It is clearly sufficient to prove that for each $f < e$, C_f puts only finitely numbers into A and B. But any number put into A or B by C_f is $x_s(f)$ for some some f; and there are only finitely many $x_s(f)$ by the induction hypothesis.

Now we show that C_{2e} is satisfied. Let x be the largest of the numbers $x_s(2e)$; we show that $x \in A$ iff $\{e\}^B(x) \simeq 1$. Since $x_s(2e)$ is increasing in s, $x_s(2e) = x$ for all sufficiently large s. If $\{e\}^B(x) \simeq 1$, then $\{e\}_s^{B^s}(x) \simeq 1$ for all large s. Choosing s this large with $(s)_0 = 2e$, we see that x is put into A. Suppose that x is put into A at step s. Since x is of the form $<2e,z>$, it is put into A by (C_{2e}); so $\{e\}_s^{B^s}(x) \simeq 1$. Hence it is enough to show that no number $< s$ is put into B after step s. Suppose that $x_t(f) < s$ is put into B at step $t > s$. Then f is odd; so $f \neq 2e$. If $2e < f$, then $x_t(f) > R_t(f) \geq r_t(2e) \geq s$, a contradiction. Thus $f < 2e$; so $R_{t+1}(2e) \geq r_{t+1}(f) \geq t > s \geq R_s(2e)$ and hence $x_{t+1}(2e) > x_s(2e) = x$, contradicting the choice of x. A similar proof shows that C_{2e+1} is satisfied. □

If $F \equiv_R G$, then by 12.4 the same functions are recursive in F and G. Hence if we have relativized some property Q, then <u>Q in F</u> is equivalent to <u>Q in G</u>. Thus if $\mathbf{a} = \text{dg } F$, we may as well say <u>$Q$ in $\mathbf{a}$</u> for <u>Q in F</u>. We define a degree $\mathbf{b}$ to be <u>RE in $\mathbf{a}$</u> if $\mathbf{b}$ contains a relation RE in $\mathbf{a}$. The relativization of the Friedberg–Muchnik Theorem tells us that for each $\mathbf{a}$, there are degrees $\mathbf{b}$ and $\mathbf{c}$

RE in **a** such that neither **b** $\leq$ **c** nor **c** $\leq$ **b**. The method used to define **0**', when relativized, shows that there is a largest degree RE in **a**. It is called the jump of **a**, and is designated by **a**'. The relativized Limit Lemma shows that a real is a limit of a recursive in **a** sequence of reals iff it has degree $\leq$ **a**'.

16. Evaluation of Degrees

We shall now show how to evaluate the degrees of certain explicitly given relations.

Let Φ be a class of relations. We say a relation R is Φ complete if R is in Φ and every relation in Φ is reducible to R (where reducible is defined before 13.3). It follows that R has the largest degree of any relation in Φ; so any two Φ complete relations have the same degree. (Caution: Some authors use complete in a somewhat different way.)

EXAMPLE. If F is total, $W_e^F(x)$ is RE in F complete; its degree is the jump of dg F. Hence any RE in F complete relation has degree (dg F)'.

The degree obtained by applying the jump n times to **0** is designated by $\mathbf{0}^n$.

16.1. PROPOSITION. For every n, there is a Σ_n^0 complete set of degree $\mathbf{0}^n$ and a Π_n^0 complete set of degree $\mathbf{0}^n$.

Proof. We use induction on n. If $n = 1$, let P be a recursive set; if $n > 1$, let P be a Π_{n-1}^0 complete set of degree $\mathbf{0}^{n-1}$. Then $W_e^P(x)$ has degree $\mathbf{0}^n$ by the example. By Post's Theorem, Σ_n^0 is the class of relation RE in P; so $W_e^P(x)$ is Σ_n^0 complete. Then $\neg W_e^P(x)$ is of degree $\mathbf{0}^n$ and is Π_n^0 complete. □

16.2. COROLLARY. Every Σ_n^0 complete or Π_n^0 complete relation has degree $\mathbf{0}^n$. □

If Φ is a class of RE sets, then the set of indices of sets in Φ is called the index set of Φ.

16.3. PROPOSITION (RICE). If Φ is a non-empty class of RE sets which is not the class of all RE sets, then the index set of Φ is not recursive.

Proof. We may suppose the empty set ϕ is not in Φ; otherwise we replace Φ be the class of RE sets which are not in Φ. Let A be an RE set which is in Φ, and let B be a non–recursive RE set. By the Parameter Theorem, there is a recursive real S such that

$$W_{S(e)}(x) \leftrightarrow x \in A \;\&\; e \in B.$$

If $e \in B$, then $W_{S(e)} = A$, so $S(e)$ is in the index set of Φ; if $e \notin B$, $W_{S(e)} = \phi$, so $S(e)$ is not in the index set of Φ. Thus B is reducible to the index set of Φ; so this index set is not recursive. □

We are going to use 16.2 to evaluate the degrees of certain index sets.

Let TOT be the index set of the class of RE sets whose only member is ω. Then

$$e \in \mathrm{TOT} \leftrightarrow \forall x W_e(x).$$

Since $W_e(x)$ is an RE relation of e and x, TOT is Π^0_2. We shall show that it is Π^0_2 complete and hence of degree $\mathbf{0}''$. Every Π^0_2 relation is reducible to its contraction, which is also Π^0_2; so it will suffice to show that every Π^0_2 set A is reducible to TOT. We have $A(x) \leftrightarrow \forall y P(x,y)$ where P is RE. By the RE Parameter Theorem, there is a recursive total S such that $W_{S(x)}(y) \leftrightarrow P(x,y)$. Hence

$$A(x) \leftrightarrow \forall y\, W_{S(x)}(y) \leftrightarrow S(x) \in \mathrm{TOT}.$$

Thus A is reducible to TOT.

Let INF be the index set of the class of infinite RE sets. Then

$$e \in \mathrm{INF} \leftrightarrow \forall x \exists y (y > x \;\&\; W_e(y)).$$

Hence INF is Π^0_2. We shall show that it is Π^0_2 complete. Let A be a Π^0_2 set. Then, writing $\mathrm{I}z$ for *for infinitely many z*,

$$A(x) \leftrightarrow \forall y P(x,y) \leftrightarrow \mathrm{I}z (\forall y < z) P(x,y)$$

where P is RE. By the table, $(\forall y < z)P(x,y)$ is an RE relation of z and x. Hence by the RE Parameter Theorem, there is a recursive total S such that

$W_{S(x)}(z) \leftrightarrow (\forall y < z)P(x,y)$. Then

$$A(x) \leftrightarrow \mathrm{I}z\, W_{S(x)}(z) \leftrightarrow S(x) \in \mathrm{INF}.$$

We say that A is reducible to B,C if there is a recursive real F such that for all x, $x \in A \to F(x) \in B$ and $x \notin A \to F(x) \notin C$. Then A is reducible to every set D such that $B \subseteq D \subseteq C$.

Let COF be the index set of the class of cofinite sets. (A set is cofinite if its complement is finite.) Since

$$e \in \mathrm{COF} \leftrightarrow \exists x \forall y(\neg W_e(y) \to y \leq x),$$

COF is Σ^0_3.

Let REC be the index set of the class of recursive sets. By 14.6,

$$e \in \mathrm{REC} \leftrightarrow \exists f(W_f = W_e^{\mathrm{c}})$$

(where a superscript c indicates a complement). Now

$$W_f = W_e^{\mathrm{c}} \leftrightarrow \forall x(W_f(x) \vee W_e(x)) \ \&\ \neg \exists x(W_f(x) \ \&\ W_e(x)).$$

The right side is Π^0_2; so REC is Σ^0_3.

Since COF $\subseteq$ REC, the following result shows that both COF and REC are Σ^0_3 complete.

16.4. Proposition (Rogers). Every Σ^0_3 set is reducible to COF,REC.

Proof. Let A be Σ^0_3. For each z, we give an RE construction of a set B_z so that B_z is cofinite if $z \in A$ and B_z is non–recursive if $z \notin A$. Moreover, we will insure that x is put into B_z at step s is a recursive relation of x,s, and z. Since

$$x \in B_z \leftrightarrow \exists s(x \text{ is put into } B_z \text{ at step } s),$$

it follows from the Parameter Theorem that there is a recursive real S such that $W_{S(z)} = B_z$ for all z. The proposition will clearly follow.

Since $z \in A \leftrightarrow \exists y P(z,y)$ where P is Π^0_2 and hence reducible to INF, there is a total recursive function F such that

$$z \in A \leftrightarrow \exists y(W_{F(z,y)} \text{ is infinite}).$$

Using 14.7, choose a one–one recursive real G such that the range of G is not recursive.

Now we describe step s in the construction of B_z. Let $B_z^{\,s}$ be the finite set of numbers put into B_z before step s, and let $x_0^{\,s}$, $x_1^{\,s}$, ... be the members of the complement of $B_z^{\,s}$ in increasing order. We put $x_{G(s)}^{\,s}$ into B_z. For each $y < s$ such that $W_{F(z,y),s+1}$ contains a number not in $W_{F(z,y),s}$, we put $x_y^{\,s}$ into B_z.

Suppose that $z \in A$. Choose y so that $W_{F(z,y)}$ is infinite. Since each $W_{F(z,y),s}$ is finite, there are infinitely many s at which $x_y^{\,s}$ is put into B_z. It follows easily that the complement of B_z has at most y members; so B_z is cofinite.

Now suppose that $z \notin A$, so that each $W_{F(z,y)}$ is finite. Since G is one–one, we see that for each y, there are only finitely many steps s at which $x_y^{\,s}$ is put into B_z. It follows that for each y, there is an x_y such that $x_y^{\,s} = x_y$ for all sufficiently large s. Hence x_0, x_1, ... are the members of the complements of B_z in increasing order. We show that B_z is not recursive by showing that the range of G is recursive in B_z. Let w be given; it is sufficient to find, using an oracle for B_z, a bound on the numbers s for which $G(s) = w$. Now if $G(s) = w$, then $x_w^{\,s}$ is put into B_z at step w; so $x_w^{\,s} \neq x_w$. Since $x_w^{\,s}$ is increasing in s, it suffices to find a stage s at which $x_w^{\,s} = x_w$. We can do this with an oracle for B_z, since the oracle enables us to compute x_w. □

The <u>index set</u> of a degree **a** is defined to be the index set of the class of RE sets having degree **a**. Thus REC is the index set of **0**. The result we have just proved is a special case of the following theorem of Yates: if **a** is RE, then the index set of **a** is Σ_3^0 in **a** complete. We shall not prove this result, which requires extensive use of the priority method.

17. Large RE Sets

Post's idea for solving his problem was to show that a sufficiently large RE set could not have degree **0'**, and then producing a non–recursive RE set which was this large. Although this idea did not solve Post's problem, it led to many interesting results, which we shall explore briefly.

We introduce our first type of large RE set. A simple set is an RE set whose complement is infinite but includes no infinite RE set. Note that a simple set is not recursive, since its complement is not RE. We shall see later that every RE degree other than **0** is the degree of a simple set; so we cannot show that a simple set cannot have degree **0'**. However, we shall prove the weaker result that a simple set cannot be RE complete. In order to do this, we first find a characterization of the RE complete sets.

A creating function for a set A is a recursive real F such that $F(e) \in A \leftrightarrow F(e) \in W_e$ for all e. A creative set is an RE set which has a creating function. For example, if A is defined by $e \in A \leftrightarrow e \in W_e$, the A is creative with creating function I_1^1. A creative set is in some sense effectively non–recursive; for the creating function F shows that A^c is not equal to any W_e and hence is not RE.

17.1. Proposition. A set is RE complete iff it is creative.

Proof. Let A be RE complete. Then there is a recursive real F such that $W_e(e) \leftrightarrow F(e) \in A$. Using the RE Parameter Theorem, pick a recursive total S so that $W_{S(e)}(x) \leftrightarrow W_e(F(x))$. Then

$$F(S(e)) \in A \leftrightarrow W_{S(e)}(S(e)) \leftrightarrow F(S(e)) \in W_e.$$

Hence $G(e) = F(S(e))$ defines a creating function for A.

Suppose that A is creative with creating function F, and let B be any RE set. Pick a total recursive S so that $W_{S(x)}(y) \leftrightarrow x \in B$. Then

$$x \in B \leftrightarrow F(S(x)) \in W_{S(x)} \leftrightarrow F(S(x)) \in A;$$

so B is reducible to A. □

17.2. Proposition. A simple set is not RE complete.

Proof. By 17.1, it is enough to show that a creative set A is not simple. Let F be a creating function for A and let g be an index of the empty set. Choose a recursive total S so that $W_{S(e)}(x) \leftrightarrow W_e(x) \vee x = F(e)$; then $W_{S(e)} = W_e \cup \{F(e)\}$. If $W_e \subseteq A^c$, then $F(e) \in W_e \leftrightarrow F(e) \in A$ shows that $F(e) \notin W_e$ and $W_{S(e)} \subseteq A^c$. Now define a recursive real G inductively by $G(0) = g$, $G(n+1) = S(G(n))$. By induction on n, $W_{G(n)} \subseteq A^c$ and $F(G(n)) \in W_{G(n+1)} - W_{G(n)}$. Thus if we set $x \in B \leftrightarrow \exists n(x \in W_{G(n)})$, then B is an infinite RE subset of A^c. Thus A is not simple. □

Now we turn to another type of large RE set. A <u>hypersimple</u> set is a coinfinite RE set A such that there is no recursive real F for which

$$\forall x \exists y(y \notin A \ \&\ x < y \leq F(x)).$$

17.3. PROPOSITION. Every hypersimple set is simple.

Proof. Let A be RE and suppose that there is an infinite RE subset B of A^c. Define an RE R by $R(x,y) \leftrightarrow y \in B \ \&\ x < y$. By the Selector Theorem, there is a recursive selector G for R. Since B is infinite, G is total; and for all x, $G(x) \notin A \ \&\ x < G(x)$. Hence A is not hypersimple. □

17.4. PROPOSITION. There is a simple set which is not hypersimple.

Proof. Define $R(x,y) \leftrightarrow y \in W_x \ \&\ y > 2x$. Then R is RE; so by the Selector Theorem, there is a recursive selector F for R. Let A be the range of F. Since $y \in A \leftrightarrow \exists x \mathcal{G}_F(x,y)$ and $\mathcal{G}_F$ is RE by the Graph Theorem, A is RE. If $F(x)$ is defined, then $F(x) > 2x$. Hence for each z, there are at most z numbers in A which are $\leq 2z$; so there is a $y \notin A$ such that $z \leq y \leq 2z$. It follows that A is coinfinite and not hypersimple. Finally, suppose that W_e is an infinite RE subset of A^c. Then $\exists y R(e,y)$; so $F(e)$ is defined and in $W_e \cap A$, a contradiction. □

To prove an analogue of 17.2 for hypersimple sets, we need some definitions. First observe that if we have an algorithm for computing A from an oracle for B, then we may use values given by the oracle to compute a number z

and then ask the oracle if $z \in B$. Imagine a simple algorithm which never does this. Then our algorithm computes a finite number of numbers; asks the oracle which of them are in B; and then decides if the input is in A. We might as well ask the oracle about all numbers less then some number z; i. e., we might as well ask the oracle for $\overline{\chi_B}(z)$.

We now put all of this into a definition. We say that A is <u>truth–table reducible</u> (abbreviated <u>tt–reducible</u>) to B if there is a recursive real F and a recursive set C such that

$$x \in A \leftrightarrow \overline{\chi_B}(F(x)) \in C$$

for all x. B is <u>truth–table complete</u> (abbreviated <u>tt–complete</u>) if B is RE and every RE set is tt–reducible to B. It is easy to see that

$$\begin{aligned} A \text{ is reducible to } B &\rightarrow A \text{ is tt–reducible to } B \\ &\rightarrow A \text{ is recursive in } B. \end{aligned}$$

Thus complete RE sets are tt–complete and tt–complete sets have degree **0**'. (Both converses are false.)

17.5. Proposition. A hypersimple set is not tt–complete.

Proof. Suppose that A is tt–complete. Choose a recursive real F and a recursive set C such that

$$W_e(e) \leftrightarrow \overline{\chi_A}(F(e)) \in C$$

for all e. Let $J_n = \{(n)_i : i < lh(n)\}$. Then every finite set is J_n for some n. Moreover, the relation $\overline{\chi_{J_n^c}}(F(x)) \notin C$ is recursive and hence RE; so there is a recursive real S such that $W_{S(n)}(x) \leftrightarrow \overline{\chi_{J_n^c}}(F(x)) \notin C$. Then

$$\overline{\chi_A}(F(S(n)) \in C \leftrightarrow W_{S(n)}(S(n)) \leftrightarrow \overline{\chi_{J_n^c}}(F(S(n))) \notin C.$$

It follows that $(\exists y < F(S(n)))(y \in A \leftrightarrow y \in J_n)$.

To show that A is not hypersimple, it will suffice to show how to compute from x a z such that $(\exists y \notin A)(x < y \leq z)$. First compute an m such that every subset of $\{0,1,\ldots,x\}$ is J_n for some $n \leq m$. Let z be the largest of the $F(S(n))$ for

$n \le m$. There is an $n \le m$ such that $J_n = \{y: y \le x \,\&\, y \notin A\}$. Hence there is a $y < F(S(n)) \le z$ such that $y \in A \leftrightarrow (y \le x \,\&\, y \notin A)$. But this clearly implies that $y \notin A$ and $x < y$. □

For the results of this section to have any interest, we must know that there are sets which are hypersimple (and hence simple).

17.6. PROPOSITION (DEKKER). If **a** is RE and **a** $\neq$ **0**, then there is a hypersimple set of degree **a**.

Proof. Let A be an RE set of degree **a**. Since **a** $\neq$ **0**, A is non–recursive and hence infinite. By 14.7, there is a one–one recursive real F with range A. Define an RE set B by

$$x \in B \leftrightarrow \exists y(x < y \,\&\, F(y) < F(x)).$$

We will show that B is hypersimple and dg $B =$ **a**.

To show that B is coinfinite, let z be given. Choose $x > z$ so that $F(x)$ is as small as possible. Then clearly $x \notin B$. Now suppose that there is a recursive real G such that for every z there is a $y \in B^C$ such that $z < y \le G(z)$. We obtain a contradiction by showing that A is recursive. Given a, $F(y) \ge a$ for large enough y (since F is one–one); so there is an z such that $a \le F(y)$ for $z < y \le G(z)$. We can find such a z by examining $z = 0$, $z = 1$, etc. We claim that $a \in A$ iff $a = F(x)$ for some $x \le G(z)$. If not, $a = F(x)$ for some $x > G(z)$. Choose $y \notin B$ such that $z < y \le G(z) < x$. Since $y \notin B$, $F(y) < F(x) = a$, contradicting the choice of z.

It remains to show that $B \equiv_R A$. Clearly

$$x \in B \leftrightarrow (\exists a < F(x))(a \in A \,\&\, (\forall y \le x)(a \neq F(y)));$$

so $B \le_R A$. Suppose that we want to use an oracle for B to compute whether or not $a \in A$. Since B is coinfinite, we can find an $x \notin B$ such that $F(x) > a$ by trial. Since $x \notin B$, $a \in A$ iff $a = F(y)$ for some $y < x$. □

17.7. COROLLARY. There is an RE set of degree **0'** which is not tt–complete.

Proof. By 17.5. □

A maximal set is a coinfinite RE set A such that for every coinfinite RE set B including A, $B - A$ is finite. Thus a maximal set is a coinfinite RE set with as few RE sets as possible including it.

It is fairly easy to show that a maximal set is hypersimple. However, it is not a simple matter to show that maximal sets exist; this was done by Friedberg. The final result of a series of investigations of this question is the following theorem of Martin: an RE degree **a** contains a maximal set iff $\mathbf{a}' = \mathbf{0}''$. Thus this notion of largeness does tell us more about the degree than our previous notions, but does not tell us that the degree cannot be $\mathbf{0}'$.

18. Function of Reals

We now extend our notion of a function to allow reals as arguments. (We could allow all total functions as arguments; but this would complicate matters without really adding anything, since a function can be replaced by its contraction.) We use lower case Greek letters, usually α, β, and γ, for reals. When the value of m is not important, we write $\vec{\alpha}$ for $\alpha_1,\ldots,\alpha_m$. We use $\mathbb{R}$ for the class of reals and $\mathbb{R}^{m,k}$ for the class of all $(m+k)$–tuples $(\alpha_1,\ldots,\alpha_m,x_1,\ldots,x_k)$. An (m,k)–ary function is a mapping of a subset of $\mathbb{R}^{m,k}$ into ω. (Thus a $(0,k)$–ary function is just a k–ary function.) From now on, a function is always an (m,k)–ary function for some m and k. Such a function is total if its domain is all of $\mathbb{R}^{m,k}$. An (m,k)–ary relation is a subset of $\mathbb{R}^{m,k}$. We define the representing function of such a relation as before.

Note that the real arguments to a function or relation must precede the number arguments. It may sometimes be convenient to write them in a different order. It is then understood that we are to move all real arguments to the left of all number arguments without otherwise changing the order of the arguments.

Now we consider how to extend the idea of computability. The new question is: how are we to be given the real inputs? The obvious answer is that we are given an oracle for each real input.

We now modify the basic machine accordingly. For each m, we have a machine called the m–real machine. (The 0–real machine will be identical with the basic machine.) In addition to the parts of the basic machine, the m–real machine has m real registers $\mathcal{F}1,\ldots \mathcal{F}m$. At any moment, each of these registers contains a real α.

We have one new type of instruction. It has the format

$$\mathcal{F}k(\mathcal{R}i) \to \mathcal{R}j$$

where $1 \leq k \leq m$ and $i \neq j$. If the machine executes this instruction when α is in $\mathcal{F}k$ and x is in $\mathcal{R}i$, it changes the number in $\mathcal{R}j$ to $\alpha(x)$ and increases the number in the counter by 1. Note that no instruction changes the contents of a real register.

With each program P and each m and k, we associate an algorithm $A_P^{m,k}$ with m real inputs and k number inputs. To perform this algorithm with the inputs $\alpha_1,\ldots,\alpha_m,x_1,\ldots,x_k$, we put P in the program holder; $\alpha_1,\ldots\alpha_m$ in $\mathcal{F}1,\ldots,\mathcal{F}m$ respectively; $x_1,\ldots,x_k$ in $\mathcal{R}1,\ldots,\mathcal{R}k$ respectively; and 0 in all other registers. We then start the machine. If the machine ever halts, the number in $\mathcal{R}0$ after it halts is the output; otherwise, there is no output. The function computed by $A_P^{m,k}$ is called the (m,k)–ary function computed by P. An (m,k)–ary function F is recursive if there is a program P such that F is the (m,k)–ary function computed by P.

We now propose to extend our previous results to these new functions. We shall have something to say only when the extensions present some problems.

We extend §4 without difficulty. The last macro now reads

$$F(\mathcal{F}p_1,\ldots,\mathcal{F}p_m,\mathcal{R}i_1,\ldots,\mathcal{R}i_k) \to \mathcal{R}j.$$

(As before, $i_1,\ldots,i_k$ should be distinct; but $p_1,\ldots,p_m$ need not be. The reader should check that this is all right.)

In §5, we need some changes to allow for the real arguments. The initial functions are now the $I_i^{m,k}$, 0, Sc, and Ap, where $I_i^{m,k}(\alpha_1,\ldots,\alpha_m,x_1,\ldots,x_k) = x_i$ and $Ap(\alpha,x) = \alpha(x)$. The program for Ap consists of the one instruction $\mathcal{F}1(\mathcal{R}1) \to \mathcal{R}0$. In the definition of closed under composition, F is now defined by

$$F(\alpha_1,\ldots,\alpha_m,\vec{x}) \simeq G(\alpha_{i_1},\ldots,\alpha_{i_r},H_1(\alpha_1,\ldots,\alpha_m,\vec{x}),\ldots,H_k(\alpha_1,\ldots,\alpha_m,\vec{x})).$$

In the case of inductively closed and μ–closed, all the functions have the same sequence $\vec{\alpha}$ of real arguments.

In §6, there are a couple of new details in the proof of 6.1. First, there is a new context $\alpha(\_\_\_\_)$. We take care of this by replacing it by $Ap(\alpha,\_\_\_\_)$. We leave it to the reader to check that our modification in the definition of closed under composition is just what is needed here.

We can extend the result in §7 that F is recursive iff $\overline{F}$ is recursive. In particular, $\overline{Ap}$ is recursive. But $\overline{Ap}(\alpha,x) = \overline{\alpha}(x)$. Thus show that $\overline{\alpha}$ (where α is a variable) is a recursive symbol.

In extending §8, we assign the code $<3,i,j,k>$ to $\mathcal{F}k(\mathcal{R}i) \to \mathcal{R}j$. Other codes are as before. (In particular, these codes do not depend in any way on what is in the real registers.) We now take $T_{k,m}(e,\alpha_1,\ldots,\alpha_m,x_1,\ldots,x_k,y)$ to mean that e is the code of a program for the real m–machine and y is the code of the P–computation from $\alpha_1,\ldots,\alpha_m,x_1,\ldots,x_k$, and leave U as before. The extension of §8 is then straightforward.

Note that if Φ is $\alpha_1,\ldots,\alpha_m$, the code of the instruction $\mathcal{F}k(\mathcal{R}i) \to \mathcal{R}j$ is the same as the code assigned to the instruction $\alpha_k(\mathcal{R}i) \to \mathcal{R}k$ for the Φ–machine; and if $\alpha_1,\ldots,\alpha_k$ are in $\mathcal{F}1,\ldots,\mathcal{F}k$, performing these two instructions has the same effect on the two machines. It follows that

$$T_{k,m}(e,\vec{\alpha},\vec{x},y) \leftrightarrow T_k^{\vec{\alpha}}(e,\vec{x},y).$$

If we use (1) of §12, we can rewrite this as

$$(1) \qquad T_{k,m}(e,\vec{\alpha},\vec{x},y) \leftrightarrow T_{k,m}(e,\vec{x},y,\overline{\vec{\alpha}}(y)).$$

(The two $T_{k,m}$'s are different, but no confusion will result.) It follows by the Normal Form Theorem that

$$\{e\}(\vec{\alpha},\vec{x}) \simeq \{e\}^{\vec{\alpha}}(\vec{x}). \tag{2}$$

The definitions and results of §12 can be extended without difficulty; but this does not give us what we really want if the functions in Φ have real arguments. The problem is that the F–instructions are not general enough. They evaluate F only at real arguments which are in a real register. Since the contents of a real register do not change during a computation, we cannot evaluate F at real arguments computed during the computation.

We shall therefore restrict the Φ in relative recursion to consists of total function of number arguments only. The results of §12 then extend without difficulty. We call the machine obtained from the real m–machine by adding the F–instructions for F in Φ the <u>Φ–m–machine</u>. If Φ is a finite sequence $H_1,\ldots,H_p$ of reals, we assign to the H_k–instruction $H_k(\mathfrak{r}i) \rightarrow \mathfrak{r}j$ the code $<3,i,j,m+k>$. In place of (2) we now have $\{e\}^{\Phi}(\vec{\alpha},\vec{x}) \simeq \{e\}^{\vec{\alpha},\Phi}(\vec{x})$. From this and (2),

$$\{e\}^{\Phi}(\vec{\alpha},\vec{x}) \simeq \{e\}(\vec{\alpha},\Phi,\vec{x}). \tag{3}$$

This gives the following alternative definition of relative recursion.

18.1. Proposition. A function F is recursive in Φ iff it has a definition $F(\vec{\alpha},\vec{x}) \simeq G(\vec{\beta},\vec{\alpha},\vec{x})$ where G is recursive and $\vec{\beta}$ is a sequence of contractions of functions in Φ. □

Remark. Even if the F in 18.1 is total, we cannot always take the G to be total.

We now consider substitution for real variables. Of course, we cannot substitute something like $F(\_\_\_\_)$ for α, since $F(\_\_\_\_)$ is a number. We therefore need some notation. We let $\lambda x(..x..)$ be the unary function F defined by $F(x) \simeq ..x..$. Then $\lambda x(..x..)$ is a real iff $..x..$ is defined for all x.

18.2. Substitution Theorem. If G and H are recursive, there is a recursive F such that

$$(4) \qquad F(\vec{\alpha},\vec{x}) \simeq G(\lambda z H(z,\vec{\alpha},\vec{x}),\ \vec{\alpha},\ \vec{x})$$

for all $\vec{\alpha},\vec{x}$ such that $\lambda z H(z,\vec{\alpha},\vec{x})$ is a real. In particular, if H is total, then the F defined by (4) is recursive.

Proof. Let g be an index of G. If $\lambda z H(z,\vec{\alpha},\vec{x})$ is a real, then the right side of (4) is, by (1),

$$U(\mu y T_{m,k}(g,\vec{x},y,\overline{H}(y,\vec{\alpha},\vec{x}),\overline{\vec{\alpha}}(y))).$$

We can use this as our definition of $F(\vec{\alpha},\vec{x})$. □

In particular, it follows that λy is a recursive expression when it is used in front of an expression defined for all values of y.

Remark. If H is not total, there may be $\vec{\alpha},\vec{x}$ such that $F(\vec{\alpha},\vec{x})$ is defined, but such that $\lambda z H(y,\vec{\alpha},\vec{x})$ is not a real and hence such that the right side of (4) is not defined.

The results of §13 and §14 extend without difficulty. However, in §13 it is natural to consider a further extension in which we allow quantifiers on real variables. We investigate this in the next section.

19. The Analytical Hierarchy

A relation is analytical if it has an explicit definition with a prefix consisting of quantifiers, which may be either universal or existential and may be on either number variables or real variables, and a recursive matrix. The basic theory of analytical relations is due to Kleene.

We begin with some rules for simplifying prefixes. As before, these may change the matrix, but they leave it recursive.

Two quantifiers are of the same kind if they are both universal or both existential; they are of the same type if they are both on real variables or both on number variables.

I. Two adjacent quantifiers of the same kind and same type can be replaced by one quantifier of that kind and type.

For number quantifiers, this is just contraction of quantifiers as in §13. In order to treat real quantifiers, we need an analogue of $(x)_i$ for reals. We define $(\alpha)_x$ to be $\lambda y\, \alpha(<x,y>)$. Here we have a better result than for numbers: given an infinite sequence $\alpha_0,\alpha_1,\ldots$ of reals, there is an α such that $(\alpha)_i = \alpha_i$ for all i. In fact, we can define α by $\alpha(x) = \alpha_{(x)_0}((x)_1)$. Moreover, $(\alpha)_x$ is a recursive expression when used in contexts of the form $(\alpha)_x($_____$)$; for we may replace this context by $\alpha(<x,$_____$>)$. We can then justify contraction of quantifiers for real variables just as we did for number variables in §13.

II. A number quantifier can be replaced by a real quantifier of the same kind.

This follows from the equivalences

$$\forall x P(x) \leftrightarrow \forall \alpha P(\alpha(0)),$$
$$\exists x P(x) \leftrightarrow \exists \alpha P(\alpha(0)).$$

III. If a number quantifier is immediately followed by a real quantifier, the real quantifier may be moved to the front of the number quantifier.

This follows from the equivalences

$$\forall x \forall \alpha P(\alpha,x) \leftrightarrow \forall \alpha \forall x P(\alpha,x),$$
$$\exists x \exists \alpha P(\alpha,x) \leftrightarrow \exists \alpha \exists x P(\alpha,x),$$
$$\forall x \exists \alpha P(\alpha,x) \leftrightarrow \exists \alpha \forall x P((\alpha)_x,x),$$
$$\exists x \forall \alpha P(\alpha,x) \leftrightarrow \forall \alpha \exists x P((\alpha)_x,x).$$

The first two of these are obvious. In the third, both sides say that there is an infinite sequence α_0, α_1, ... such that $P(\alpha_x,x)$ for all x. If we put $\neg P$ for P in the third, bring the quantifiers to the front by prenex rules, and drop the $\neg$ from both sides, we get the fourth.

We say that a prefix is $\underline{\Pi}_n^1$ ($\underline{\Sigma}_n^1$) if all of the real quantifiers precede all of the number quantifiers; there are exactly n real quantifiers; the real quantifiers alternate in kind; and the prefix begins with $\forall$ ($\exists$). A relation is $\underline{\Pi}_n^1$ if it has a definition with a Π_n^1 prefix and a recursive matrix; similarly for Σ_n^1. A relation is $\underline{\Delta}_n^1$ if it is both Π_n^1 and Σ_n^1.

19.1. Proposition. Every analytical relation is either Π_n^1 or Σ_n^1 for some n.

Proof. By first using III, then applying contraction of quantifiers to the real variables. □

A Π_n^1 or Σ_n^1 prefix may be further simplified as follows. If there are any number quantifiers of the same kind as the last real quantifier, we change then to real quantifiers by II; move then to just after the last real quantifier by III; and then contract them with the last real quantifier. The remaining number quantifiers are of the opposite kind to the last real quantifier, and may be contracted to one number quantifier of this kind. If there are no number quantifiers left, we can add a superfluous one of the opposite kind to the last real quantifier. In summary: we may suppose that there is exactly one number quantifier, which is if the opposite kind to the last real quantifier.

We can still say some more. Consider, for example, a Π_1^1 relation P. By the above, we have $P(\vec{\alpha},\vec{x}) \leftrightarrow \forall\beta Q(\vec{\alpha},\vec{x},\beta)$ where Q is Σ_1^0. Then, using (3) of §14, $P(\vec{\alpha},\vec{x}) \leftrightarrow \forall\beta\exists y R(\overline{\vec{\alpha}}(y),\vec{x},\overline{\beta}(y),y)$ with R recursive. We can even omit the last y, since it may be replaced by $\mathrm{lh}(\overline{\beta}(y))$. Thus any Π_1^1 relation of $\vec{\alpha},\vec{x}$ can be written $\forall\beta\exists y R(\overline{\vec{\alpha}}(y),\vec{x},\overline{\beta}(y))$ with R recursive. Taking negations, any Σ_1^1 relation of $\vec{\alpha},\vec{x}$ can be written $\exists\beta\forall y R(\overline{\vec{\alpha}}(y),\vec{x},\overline{\beta}(y))$ with R recursive. Similar results hold for Π_k^1 and Σ_k^1 relations.

19.2. Proposition. If R is Π_n^1 or Σ_n^1, then R is Δ_k^1 for all $k > n$. If R is arithmetical, it is Δ_n^1 for all n.

Proof. By adding superfluous quantifiers. □

We shall now say that P is reducible to Q if

$$P(\vec{\alpha},\vec{x}) \leftrightarrow Q(\lambda y G_1(y,\vec{\alpha},\vec{x}),\ldots,\lambda y G_m(y,\vec{\alpha},\vec{x}),F_1(\vec{\alpha},\vec{x}),\ldots,F_k(\vec{\alpha},\vec{x}))$$

where $G_1,\ldots,G_m,F_1,\ldots,F_k$ are total and recursive.

19.3. PROPOSITION. If P is Π^1_n and Q is reducible to P, then P is Π^1_n; and similarly with Σ^1_n or Δ^1_n in place of Π^1_n. □

The analogue of the table in §12 is the following table.

P,Q	$\neg P$	$P \vee Q$	$P \,\&\, Q$	$\forall\alpha P$	$\exists\alpha P$	$Qx P$
Π^1_n	Σ^1_n	Π^1_n	Π^1_n	Π^1_n	Σ^1_{n+1}	Π^1_n
Σ^1_n	Π^1_n	Σ^1_n	Σ^1_n	Π^1_{n+1}	Σ^1_n	Σ^1_n
Δ^1_n	Δ^1_n	Δ^1_n	Δ^1_n	Π^1_n	Σ^1_n	Δ^1_n

It is proved and used in the same way as the earlier table.

The classification of analytical relations into the Π^1_n and Σ^1_n relations is called the analytical hierarchy.

19.4. ANALYTICAL ENUMERATION THEOREM. For every n, m, and k, there is a Π^1_n $(m,k+1)$–ary function which enumerates the class of Π^1_n (m,k)–ary relations; and similarly with Σ^1_n for Π^1_n.

Proof. Suppose, for example,we want to enumerate the Π^1_2 (1,1)–ary relations. Every such relation R is of the form $\forall\alpha\exists\beta P$ where P is Π^0_1 by the remarks after 19.1. Thus if Q is Π^0_1 and enumerates the Π^0_1 (3,1)–ary relations, then $\forall\alpha\exists\beta Q(\alpha,\beta,\gamma,x,e)$ is the desired enumerating function. □

19.5. ANALYTICAL HIERARCHY THEOREM. For each n, there is a Π^1_n set which is not Σ^1_n, hence not Π^1_k or Σ^1_k for any $k < n$. The same holds with Π^1_n and Σ^1_n interchanged.

Proof. As in the arithmetical case. □

20. The Projective Hierarchy

The results of the last section can be relativized to a class Φ of total functions of number variables. A particularly interesting case is that in which Φ

is the class of all such functions. Replacing the functions by their contractions, we see we are relativizing to the class $\mathbb{R}$ of reals. Note that by 18.1, a function is recursive in $\mathbb{R}$ iff it is obtained from a recursive function by replacing some of the unary function variables by names of particular reals. The same then holds with *recursive* replaced by Π^1_k or Σ^1_k.

A relation is <u>projective</u> if it is analytical in $\mathbb{R}$. The analytical hierarchy relativized to $\mathbb{R}$ is called the <u>projective hierarchy</u>. (It is customary to write a boldface Π^1_n for Π^1_n in $\mathbb{R}$ and similarly for Σ and Δ. We avoid this notation, since boldface is sometimes hard to distinguish from lightface.) The theory of the projective hierarchy antedates that of the analylytical hierearchy; it was begun by Lusin, Suslin, and Sierpinski.

The Enumeration Theorem does not hold in its usual form for Π^1_n in $\mathbb{R}$; but we shall prove a modified form. We say that a $(m+1,k)$–ary relation Q <u>$\mathbb{R}$–enumerates</u> a class Φ of (m,k)–ary relations if for every R in Φ, there is a β such that $R(\vec{\alpha},\vec{x}) \leftrightarrow Q(\vec{\alpha},\vec{x},\beta)$ for all $\vec{\alpha}$ and $\vec{x}$.

20.1. PROJECTIVE ENUMERATION THEOREM. For every n, m, and k, there is a $(m+1,k)$–ary Π^1_n relation which $\mathbb{R}$–enumerates the class of (m,k)–ary Π^1_n in $\mathbb{R}$ relations; and similarly with Σ^1_n for Π^1_n.

Proof. As in the proof of the analytical case, it is enough to do this for Σ^0_1, i.e., RE. If R is RE in $\mathbb{R}$, it is RE in a finite sequence $\vec{\alpha}$ of reals. If e is a $\vec{\alpha}$–index of R, then by (3) of §18,

$$\begin{aligned} R(\vec{\beta},\vec{x}) &\leftrightarrow \{e\}^{\vec{\alpha}}(\vec{\beta},\vec{x}) \text{ is defined} \\ &\leftrightarrow \{e\}(\vec{\beta},\vec{\alpha},\vec{x}) \text{ is defined} \\ &\leftrightarrow W_e(\vec{\beta},\vec{\alpha},\vec{x}). \end{aligned}$$

Choose γ so that $(\gamma)_0(0) = e$ and $(\gamma)_i = \alpha_i$ for $1 \le i \le n$. Then the right side becomes $W_{(\gamma)_0(0)}(\vec{\beta},(\gamma)_1,\ldots,(\gamma)_m,\vec{x})$. This is an RE relation P of $\vec{\beta},\vec{x},\gamma$; and P is the desired enumerating relation. □

We leave it to the reader to derive a Projective Hierarchy Theorem from this; the examples will now be (1,0)–ary. (Every $(0,k)$–ary relation is recursive in $\mathbb{R}$.)

20.2. PROPOSITION. Let P be defined by $P(\vec{\alpha},\vec{x},y) \leftrightarrow P_y(\vec{\alpha},\vec{x})$. Then P is Π^1_n in $\mathbb{R}$ iff each P_y is Π^1_n in $\mathbb{R}$; and similarly with Σ^1_n or Δ^1_n in place of Π^1_n.

Proof. If P is Π^1_n in $\mathbb{R}$, each P_y is clearly Π^1_n in $\mathbb{R}$. Now suppose that each P_y is Π^1_n in $\mathbb{R}$. By the Projective Enumeration Theorem, there is a Π^1_n relation Q and a β_y for each y such that $P_y(\vec{\alpha},\vec{x}) \leftrightarrow Q(\vec{\alpha},\vec{x},\beta_y)$. Choose β so that $(\beta)_y = \beta_y$ for all y. Then $P(\vec{\alpha},\vec{x},y) \leftrightarrow Q(\vec{\alpha},\vec{x},(\beta)_y)$. Thus P is Π^1_n in β and hence in $\mathbb{R}$. □

The further study of the analytical and projective hierarchies is known as Descriptive Set Theory, and is a hybrid of Recursion Theory and Set Theory. We shall prove only one result. We shall prove it for the projective hierarchy; the analogue for the analytical hierarchy is more difficult both to state and to prove.

We recall a definition from measure theory. Let X be a space and let Λ be a class of subsets of X. We say that Λ is a <u>σ–ring</u> if: (a) the complement of every set in Λ is in Λ; (b) every countable union of sets in Λ is in Λ; (c) $X \in \Lambda$. From (a) and (b) it follows that: (d) every countable intersection of sets in Λ is in Λ. If Γ is any collection of subsets of X, there is a smallest σ–ring including Γ; it is the intersection of all of the Σ–rings which include Γ.

20.3 PROPOSITION. The class of Δ^1_n in $\mathbb{R}$ (m,k)–ary relations is a σ–ring in $\mathbb{R}^{m,k}$.

Proof. In view of the table, it is enough to show that the union Q of a sequence $\{P_j\}$ of such relations is Δ^1_n in $\mathbb{R}$. Defining $P(\vec{\alpha},\vec{x},j) \leftrightarrow P_j(\vec{\alpha},\vec{x})$, P is Δ^1_n by 20.2. Since $Q(\vec{\alpha},\vec{x}) \leftrightarrow \exists j P(\vec{\alpha},\vec{x},j)$, Q is Δ^1_n in $\mathbb{R}$ by the table. □

An (m,k)–ary relation is <u>Borel</u> if it belongs to the smallest σ–ring in $\mathbb{R}^{m,k}$ which contains all the recursive (m,k)–ary relations. By 20.3, every Borel relation is Δ^1_1 in $\mathbb{R}$. We shall prove that the converse also holds.

Let A and B be subsets of a space X. We say that a subset C of X separates A and B if $A \subseteq C$ and $B \subseteq C^c$. This clearly implies that A and B are disjoint.

20.4. SEPARATION THEOREM. Any two disjoint Σ^1_1 in $\mathbb{R}$ (m,k)–ary relations can be separated by a Borel relation.

Proof. To make the notation simpler, let $m = 1$ and $k = 0$. Say that A is inseparable from B if no Borel relation separates A and B. We shall first prove the following lemma: If $\cup_{i\in\omega}A_i$ is inseparable from $\cup_{j\in\omega}B_j$, then there are i and j such that A_i is inseparable from B_j. Suppose, on the contrary, that for every i and j, there is a Borel relation $C_{i,j}$ which separates A_i and B_j. If $C = \cap_{j\in\omega}\cup_{i\in\omega}C_{i,j}$, then C is Borel and separates $\cup_{i\in\omega}A_i$ and $\cup_{j\in\omega}B_j$.

Now assume that P and Q are inseparable Σ^1_1 in $\mathbb{R}$ relations; we shall show that P and Q are not disjoint. Using the remarks after 19.1, we can write

$$P(\alpha) \leftrightarrow \exists\beta\forall nR(\overline{\alpha}(n),\overline{\beta}(n)),$$

$$Q(\alpha) \leftrightarrow \exists\gamma\forall nR'(\overline{\alpha}(n),\overline{\gamma}(n)),$$

where R and R' are recursive in $\mathbb{R}$. For $z,w \in Seq$, let

$$P_{z,w}(\alpha) \leftrightarrow z = \overline{\alpha}(lh(z)) \;\&\; \exists\beta(w = \overline{\beta}(lh(w)) \;\&\; \forall nR(\overline{\alpha}(n),\overline{\beta}(n))),$$

and define $Q_{z,w}$ similarly but with R replaced by R'. It is clear that

$$P_{z,w} = \cup_{m\in\omega}\cup_{p\in\omega}P_{z*<m>,w*<p>}$$

and similarly for $Q_{z,w}$.

We shall define $\alpha(n)$, $\beta(n)$, and $\gamma(n)$ by induction on n so that $P_{\overline{\alpha}(n),\overline{\beta}(n)}$ and $Q_{\overline{\alpha}(n),\overline{\gamma}(n)}$ are inseparable. Since $P = P_{<>,<>}$ and $Q = Q_{<>,<>}$, this holds for $n = 0$. Suppose it holds for some n. By our lemma, there are i, j, k, and l so that $P_{\overline{\alpha}(n)*<i>,\overline{\beta}(n)*<j>}$ is inseparable from $Q_{\overline{\alpha}(n)*<k>,\overline{\gamma}(n)*<l>}$. Then $i = k$; for otherwise, $\{\delta: \overline{\delta}(n+1) = \overline{\alpha}(n)*<i>\}$ is a recursive (and hence Borel) set which separates $P_{\overline{\alpha}(n)*<i>,\overline{\beta}(n)*<j>}$ from $Q_{\overline{\alpha}(n)*<k>,\overline{\gamma}(n)*<l>}$.

Thus we may take $\alpha(n) = i$, $\beta(n) = j$, and $\gamma(n) = l$.

For each n, $P_{\overline{\alpha}(n),\overline{\beta}(n)}$ is inseparable from $Q_{\overline{\alpha}(n),\overline{\gamma}(n)}$; so, they are both non–empty. This implies that $R(\overline{\alpha}(n),\overline{\beta}(n))$ and $R'(\overline{\alpha}(n),\overline{\gamma}(n))$ for all n. Hence $P(\alpha)$ and $Q(\alpha)$; so P and Q are not disjoint. □

20.5. SUSLIN'S THEOREM. A relation is Borel iff it is Δ^1_1 in $\mathbb{R}$.

Proof. We have already seen that every Borel relation is Δ^1_1 in $\mathbb{R}$. Now let P be Δ^1_1 in $\mathbb{R}$. Then P and $\neg P$ are Σ^1_1 in $\mathbb{R}$; so by the Separation Theorem, there is a Δ^1_1 in $\mathbb{R}$ relation which separates P and $\neg P$. But the only relation which separates P and $\neg P$ is P. □

Suggestions for Further Reading

This is not a bibliography, but a very personal selection of useful books and articles.

There are many undergraduate texts in recursion theory. My favorite is *Computability, Complexity, and Languages* by Davis and Weyuker. It treats several topics of particular interest in computer science which are not touched on in this notes.

The standard graduate text in recursion theory has always been *Theory of Recursive Functions and Effective Computability* by Rogers. Although badly out of date, it is still a valuable reference. It has a very large supply of excellent problems.

For readers with at least a little background in recursion theory, *Recursively Enumerable Sets and Degrees* by Soare is a valuable book, either for learning or for reference. It contains much more material on the topics considered in sections 14–17 of these notes, as well as material on the lattice of RE sets, which we have not touched on. It is written in a rather compressed style; the reader is expected to do his share of the work.

The standard text on Descriptive Set Theory (see §20) is *Descriptive Set Theory* by Moschovakis. Among other things, it contains much interesting historical information.

The *Handbook of Mathematical Logic*, edited by Barwise, contains valuable introductory articles on several topics in recursion theory.

There are extensive bibliographies of journal articles in the books mentioned above.

LECTURE NOTES IN LOGIC

General Remarks

This series is intended to serve researchers, teachers, and students in the field of symbolic logic, broadly interpreted. The aim of the series is to bring publications to the logic community with the least possible delay and to provide rapid dissemination of the latest research. Scientific quality is the overriding criterion by which submissions are evaluated.

Books in the Lecture Notes in Logic series are printed by photo-offset from master copy prepared using LaTeX and the ASL style files. For this purpose the Association for Symbolic Logic provides technical instructions to authors. Careful preparation of manuscripts will help keep production time short, reduce costs, and ensure quality of appearance of the finished book. Authors receive 50 free copies of their book. No royalty is paid on LNL volumes.

Commitment to publish may be made by letter of intent rather than by signing a formal contract, at the discretion of the ASL Publisher. The Association for Symbolic Logic secures the copyright for each volume.

The editors prefer email contact and encourage electronic submissions.

Editorial Board

David Marker, Managing Editor
Dept. of Mathematics, Statistics, and Computer Science (M/C 249)
University of Illinois at Chicago
851 S. Morgan St.
Chicago, IL 60607-7045
marker@math.uic.edu

Vladimir Kanovei
Lab 6
Institute for Information Transmission Problems
Bol. Karetnyj Per. 19
Moscow 127994 Russia
kanovei@mccme.ru

Steffen Lempp
Department of Mathematics
University of Wisconsin
480 Lincoln Avenue
Madison, Wisconsin 53706-1388
lempp@math.wisc.edu

Lance Fortnow
Department of Computer Science
University of Chicago
1100 East 58th Street
Chicago, Illinois 60637
fortnow@cs.uchicago.edu

Shaughan Lavine
Department of Philosophy
The University of Arizona
P.O. Box 210027
Tuscon, Arizona 85721-0027
shaughan@ns.arizona.edu

Anand Pillay
Department of Mathematics
University of Illinois
1409 West Green Street
Urbana, Illinois 61801
pillay@math.uiuc.edu

Editorial Policy

1. Submissions are invited in the following categories:

i) Research monographs
ii) Lecture and seminar notes
iii) Reports of meetings
iv) Texts which are out of print

Those considering a project which might be suitable for the series are strongly advised to contact the publisher or the series editors at an early stage.

2. Categories i) and ii). These categories will be emphasized by Lecture Notes in Logic and are normally reserved for works written by one or two authors. The goal is to report new developments quickly, informally, and in a way that will make them accessible to non-specialists. Books in these categories should include

– at least 100 pages of text;
– a table of contents and a subject index;
– an informative introduction, perhaps with some historical remarks, which should be accessible to readers unfamiliar with the topic treated;

In the evaluation of submissions, timeliness of the work is an important criterion. Texts should be well-rounded and reasonably self-contained. In most cases the work will contain results of others as well as those of the authors. In each case, the author(s) should provide sufficient motivation, examples, and applications. Ph.D. theses will be suitable for this series only when they are of exceptional interest and of high expository quality.

Proposals in these categories should be submitted (preferably in duplicate) to one of the series editors, and will be refereed. A provisional judgment on the acceptability of a project can be based on partial information about the work: a first draft, or a detailed outline describing the contents of each chapter, the estimated length, a bibliography, and one or two sample chapters. A final decision whether to accept will rest on an evaluation of the completed work.

3. Category iii). Reports of meetings will be considered for publication provided that they are of lasting interest. In exceptional cases, other multi-authored volumes may be considered in this category. One or more expert participant(s) will act as the scientific editor(s) of the volume. They select the papers which are suitable for inclusion and have them individually refereed as for a journal. Organizers should contact the Managing Editor of Lecture Notes in Logic in the early planning stages.

4. Category iv). This category provides an avenue to provide out-of-print books that are still in demand to a new generation of logicians.

5. Format. Works in English are preferred. After the manuscript is accepted in its final form, an electronic copy in LaTeX format will be appreciated and will advance considerably the publication date of the book. Authors are strongly urged to seek typesetting instructions from the Association for Symbolic Logic at an early stage of manuscript preparation.

Lecture Notes in Logic

From 1993 to 1999 this series was published under an agreement between the Association for Symbolic Logic and Springer-Verlag. Since 1999 the ASL is Publisher and A K Peters, Ltd. is Co-publisher. The ASL is committed to keeping all books in the series in print.

Current information may be found at `http://www.aslonline.org`, the ASL Web site. Editorial and submission policies and the list of Editors may also be found above.

Previously published books in the *Lecture Notes in Logic* are:

1. *Recursion Theory.* J. R. Shoenfield. (1993, reprinted 2001; 84 pp.)
2. *Logic Colloquium '90; Proceedings of the Annual European Summer Meeting of the Association for Symbolic Logic, held in Helsinki, Finland, July 15–22, 1990.* Eds. J. Oikkonen and J. Väänänen. (1993, reprinted 2001; 305 pp.)
3. *Fine Structure and Iteration Trees.* W. Mitchell and J. Steel. (1994; 130 pp.)
4. *Descriptive Set Theory and Forcing: How to Prove Theorems about Borel Sets the Hard Way.* A. W. Miller. (1995; 130 pp.)
5. *Model Theory of Fields.* D. Marker, M. Messmer, and A. Pillay. (First edition, 1996, 154 pp. Second edition, 2006, 155 pp.)
6. *Gödel '96; Logical Foundations of Mathematics, Computer Science and Physics; Kurt Gödel's Legacy. Brno, Czech Republic, August 1996, Proceedings.* Ed. P. Hajek. (1996, reprinted 2001; 322 pp.)
7. *A General Algebraic Semantics for Sentential Objects.* J. M. Font and R. Jansana. (1996; 135 pp.)
8. *The Core Model Iterability Problem.* J. Steel. (1997; 112 pp.)
9. *Bounded Variable Logics and Counting.* M. Otto. (1997; 183 pp.)
10. *Aspects of Incompleteness.* P. Lindstrom. (First edition, 1997. Second edition, 2003, 163 pp.)
11. *Logic Colloquium '95; Proceedings of the Annual European Summer Meeting of the Association for Symbolic Logic, held in Haifa, Israel, August 9–18, 1995.* Eds. J. A. Makowsky and E. V. Ravve. (1998; 364 pp.)
12. *Logic Colloquium '96; Proceedings of the Colloquium held in San Sebastian, Spain, July 9–15, 1996.* Eds. J. M. Larrazabal, D. Lascar, and G. Mints. (1998; 268 pp.)
13. *Logic Colloquium '98; Proceedings of the Annual European Summer Meeting of the Association for Symbolic Logic, held in Prague, Czech Republic, August 9–15, 1998.* Eds. S. R. Buss, P. Hájek, and P. Pudlák. (2000; 541 pp.)
14. *Model Theory of Stochastic Processes.* S. Fajardo and H. J. Keisler. (2002; 136 pp.)

15. *Reflections on the Foundations of Mathematics; Essays in Honor of Solomon Feferman.* Eds. W. Seig, R. Sommer, and C. Talcott. (2002; 444 pp.)

16. *Inexhaustibility; A Non-exhaustive Treatment.* T. Franzén. (2004; 255 pp.)

17. *Logic Colloquium '99; Proceedings of the Annual European Summer Meeting of the Association for Symbolic Logic, held in Utrecht, Netherlands, August 1–6, 1999.* Eds. J. van Eijck, V. van Oostrom, and A. Visser. (2004; 208 pp.)

18. *The Notre Dame Lectures.* Ed. P. Cholak. (2005, 185 pp.)

19. *Logic Colloquium 2000; Proceedings of the Annual European Summer Meeting of the Association for Symbolic Logic, held in Paris, France, July 23–31, 2000.* Eds. R. Cori, A. Razborov, S. Todorčević, and C. Wood. (2005; 408 pp.)

20. *Logic Colloquium '01; Proceedings of the Annual European Summer Meeting of the Association for Symbolic Logic, held in Vienna, Austria, August 1–6, 2001.* Eds. M. Baaz, S. Friedman, and J. Krajíček. (2005, 486 pp.)

21. *Reverse Mathematics 2001.* Ed. S. Simpson. (2005, 401 pp.)

22. *Intensionality.* Ed. R. Kahle. (2005, 265 pp.)

23. *Logicism Renewed: Logical Foundations for Mathematics and Computer Science.* P. Gilmore. (2005, 230 pp.)

24. *Logic Colloquium '03; Proceedings of the Annual European Summer Meeting of the Association for Symbolic Logic, held in Helsinki, Finland, August 14–20, 2003.* Eds. V. Stoltenberg-Hansen and J. Väänänen. (2006; 407 pp.)

25. *Nonstandard Methods and Applications in Mathematics.* Eds. N.J. Cutland, M. Di Nasso, and D. Ross. (2006; 248 pp.)

26. *Logic in Tehran: Proceedings of the Workshop and Conference on Logic, Algebra, and Arithmetic, held October 18–22, 2003.* Eds. A. Enayat, I. Kalantari, M. Moniri. (2006; 341 pp.)

Milton Keynes UK
Ingram Content Group UK Ltd.
UKHW031152141024
449569UK00024B/851

9 781568 811499